Math Challenge II-A
Algebra

Areteem Institute

Math Challenge II-A Algebra

Edited by   John Lensmire
David Reynoso
Kevin Wang
Kelly Ren

Copyright © 2018 ARETEEM INSTITUTE

WWW.ARETEEM.ORG

PUBLISHED BY ARETEEM PRESS

ISBN: 1-944863-24-9
ISBN-13: 978-1-944863-24-1

First printing, October 2018.

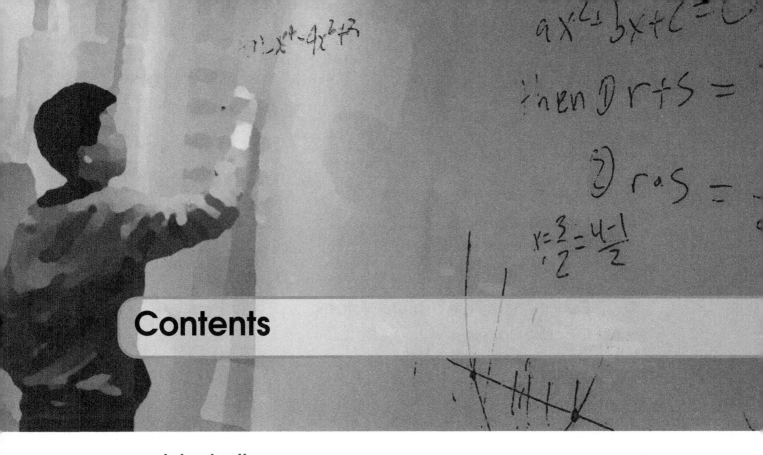

Contents

Introduction

Math Challenge II-A is for students who are preparing for the AMC 10 contest. Students are required to have fundamental knowledge in Algebra I, Geometry, Basic Number Theory and Counting and Probability up to the 10th grade level. Topics include polynomials, inequalities, special algebraic techniques, triangles and polygons, collinearity and concurrency, vectors and coordinates, numbers and divisibility, modular arithmetic, advanced counting strategies, binomial coefficients, sequence and series, and various other topics and problem solving techniques involved in math contests such as the AMC 10, advanced MathCounts, ARML, and ZIML.

The course is divided into four terms:

- Summer, covering Algebra
- Fall, covering Geometry
- Winter, covering Combinatorics
- Spring, covering Number Theory

The book contains course materials for Math Challenge II-A: Algebra.

We recommend that students take all four terms, but the terms do not build on previous terms, so they do not need to be taken in order and students can take single terms if they want to focus on specific topics.

Students can sign up for the course at `classes.areteem.org` for the live online version or at `edurila.com` for the self-paced version.

About Areteem Institute

Areteem Institute is an educational institution that develops and provides in-depth and advanced math and science programs for K-12 (Elementary School, Middle School, and High School) students and teachers. Areteem programs are accredited supplementary programs by the Western Association of Schools and Colleges (WASC). Students may attend the Areteem Institute in one or more of the following options:

- Live and real-time face-to-face online classes with audio, video, interactive online whiteboard, and text chatting capabilities;
- Self-paced classes by watching the recordings of the live classes;
- Short video courses for trending math, science, technology, engineering, English, and social studies topics;
- Summer Intensive Camps held on prestigious university campuses and Winter Boot Camps;
- Practice with selected free daily problems and monthly ZIML competitions at ziml.areteem.org.

Areteem courses are designed and developed by educational experts and industry professionals to bring real world applications into STEM education. The programs are ideal for students who wish to build their mathematical strength in order to excel academically and eventually win in Math Competitions (AMC, AIME, USAMO, IMO, ARML, MathCounts, Math Olympiad, ZIML, and other math leagues and tournaments, etc.), Science Fairs (County Science Fairs, State Science Fairs, national programs like Intel Science and Engineering Fair, etc.) and Science Olympiads, or for students who purely want to enrich their academic lives by taking more challenging courses and developing outstanding analytical, logical, and creative problem solving skills.

Since 2004 Areteem Institute has been teaching with methodology that is highly promoted by the new Common Core State Standards: stressing the conceptual level understanding of the math concepts, problem solving techniques, and solving problems with real world applications. With the guidance from experienced and passionate professors, students are motivated to explore concepts deeper by identifying an interesting problem, researching it, analyzing it, and using a critical thinking approach to come up with multiple solutions.

Thousands of math students who have been trained at Areteem have achieved top honors and earned top awards in major national and international math competitions, including Gold Medalists in the International Math Olympiad (IMO), top winners and qualifiers at the USA Math Olympiad (USAMO/JMO) and AIME, top winners at the

Zoom International Math League (ZIML), and top winners at the MathCounts National Competition. Many Areteem Alumni have graduated from high school and gone on to enter their dream colleges such as MIT, Cal Tech, Harvard, Stanford, Yale, Princeton, U Penn, Harvey Mudd College, UC Berkeley, or UCLA. Those who have graduated from colleges are now playing important roles in their fields of endeavor.

Further information about Areteem Institute, as well as updates and errata of this book, can be found online at http://www.areteem.org.

Acknowledgments

This book contains many years of collaborative work by the staff of Areteem Institute. This book could not have existed without their efforts. Huge thanks go to the Areteem staff for their contributions!

The examples and problems in this book were either created by the Areteem staff or adapted from various sources, including other books and online resources. Especially, some good problems from previous math competitions and contests such as AMC, AIME, ARML, MATHCOUNTS, and ZIML are chosen as examples to illustrate concepts or problem-solving techniques. The original resources are credited whenever possible. However, it is not practical to list all such resources. We extend our gratitude to the original authors of all these resources.

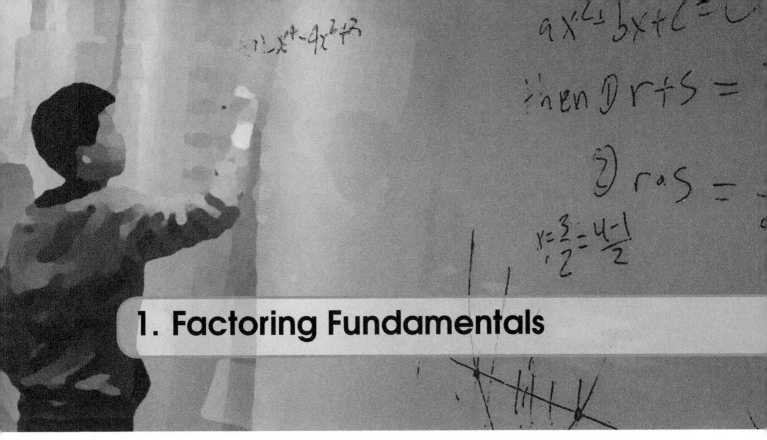

1. Factoring Fundamentals

Factoring Basics

- **Finding Common Factors**: The first step in any factoring problem is to search for common factors. Factoring out the common factor is the backward application of the Distributive Law: $ab + ac = a(b + c)$. Here a, b, c can be numbers or expressions. Sometimes it takes some work in order to get the common factor to appear.

- **Commonly used formulas**:

$$a^2 - b^2 = (a + b)(a - b)$$
$$a^2 \pm 2ab + b^2 = (a \pm b)^2$$
$$a^3 \pm 3a^2b + 3ab^2 \pm b^3 = (a \pm b)^3$$
$$a^3 + b^3 = (a + b)(a^2 - ab + b^2)$$
$$a^3 - b^3 = (a - b)(a^2 + ab + b^2)$$
$$a^2 + b^2 + c^2 + 2ab + 2bc + 2ca = (a + b + c)^2$$
$$a^3 + b^3 + c^3 - 3abc = (a + b + c)(a^2 + b^2 + c^2 - ab - bc - ca)$$
$$a^n - b^n = (a - b)(a^{n-1} + a^{n-2}b + a^{n-3}b^2 + \cdots + ab^{n-2} + b^{n-1}), \quad n \in \mathbb{N}$$
$$a^n - b^n = (a + b)(a^{n-1} - a^{n-2}b + a^{n-3}b^2 - \cdots + ab^{n-2} - b^{n-1}), \quad n \in \mathbb{N} \text{ even}$$
$$a^n + b^n = (a + b)(a^{n-1} - a^{n-2}b + a^{n-3}b^2 - \cdots - ab^{n-2} + b^{n-1}), \quad n \in \mathbb{N} \text{ odd}$$

- **Cross-multiplication**: Cross-multiplication is the reverse of the FOIL method: $abx^2 + (ad + bc)x + cd = (ax + c)(bx + d)$.

Further Factoring Techniques

- **Grouping**: If there are 4 or more terms, we can group some terms together and

---

factor the portions first, and sometimes the portions end up with common factors or fit into a formula, then factoring is possible.

- **Split and add terms**: Split one term to two or more terms; add two terms that cancel each other. These techniques can enable the polynomial to be factored by grouping.

1.1   Example Questions

Problem 1.1 Factor the following.

(a) $10x^{10}y^8 + 5x^5y^9$.

(b) $(2x+3y)^3 - 8x^3 - 27y^3$.

Problem 1.2 Prove the following identities:

(a) Prove the formula for factoring $a^3 + b^3 + c^3 - 3abc$.

(b) Prove the formula for factoring $y^6 - y^3$.

Problem 1.3 Factor $x^{15} + x^{14} + x^{13} + \cdots + x^2 + x + 1$.

Problem 1.4 Factor the following.

(a) Factor $22y^2 - 35y + 3$.

(b) Factor $2x^2 - 7x + 3$.

(c) Factor $3k^2 - 5k - 2$.

Problem 1.5 Factor $a^4 + a^3 + a^2b + ab^2 + b^3 - b^4$.

Problem 1.6 Factor $x^3 - 9x + 8$ by using the given technique:

(a) $8 = -1 + 9$.

(b) $-9x = -x - 8x$.

(c) $x^3 = 9x^3 - 8x^3$.

(d) Add two terms: $-x^2 + x^2$.

Problem 1.7 Factor the following.

(a) $x^9 + x^6 + x^3 - 3$.

(b) (Sophie Germain's technique) $x^4 + 4y^4$.

Problem 1.8 Factor the expression $(x+1)^4 + (x^2-1)^2 + (x-1)^4$.

Problem 1.9 Factor $a^2 + b^2 + c^2 - 2bc + 2ca - 2ab$.

Problem 1.10 Factor $-2x^{5n-1}y^n + 4x^{3n-1}y^{n+2} - 2x^{n-1}y^{n+4}$.

1.2 Quick Response Questions

Problem 1.11 What is the greatest common factor of the terms in $26x^3yz^3 + 91x^2y^7z^5$?

(A) $91x^3y^7z^5$
(B) $13x^2yz^3$
(C) $117x^5y^8z^8$
(D) $2x + 7y^6z^2$

Problem 1.12 Factor $1 + x + y + xy$.

(A) $x(1+y)$
(B) $(x+y)(x+y)$
(C) $(1+x)(1+y)$
(D) $(y+x)(x+1)$

Problem 1.13 $x^2 + 2xy + y^2 - 9$ factors as $(x+y-k)(x+y+k)$. What is k?

Problem 1.14 Factor $x^3 + x^2 - 4x - 4$.

(A) $x^2(x-4)$
(B) $(x-2)(x+1)(x-1)$
(C) $x^3(1-4x)$
(D) $(x+2)(x-2)(x+1)$

Problem 1.15 Factor $1 + x + y + z + xy + xz + yz + xyz$.

(A) $(1+x)(1+y)(1+z)$
(B) $z(1+x+y+xy)$
(C) $(x+y+z)(1+xyz)$
(D) $(1+x)(y+z)$

Problem 1.16 Factor $x^2 - 6x + 8$ using $8 = 9 - 1$.

(A) $(x+9)(x-1)$
(B) $(x+3)(x-3-1)$
(C) $(x-4)(x-2)$
(D) $(x+4)(x-2)$

Problem 1.17 Factor $2x^2 + 7x + 6$.

(A) $(2x+1)(x+3)$
(B) $(2x+3)(x+2)$
(C) $(x+3)(2x+2)$
(D) $(2x+1)(x+6)$

Problem 1.18 Factor $6x^2 + x - 35$.

(A) $(2x-5)(3x+7)$
(B) $(6x+1)(x-35)$
(C) $(2x+5)(3x-7)$
(D) $(3x+5)(2x-7)$

Problem 1.19 Factor $x^4 - 5x^2 + 4$.

(A) $(x^2 + 5)(x^2 - 1)$
(B) $(x - 1)(x + 1)(x - 2)(x + 2)$
(C) $(x - 1)(x - 1)(x + 2)(x + 2)$
(D) $(x^2 + 5)(x + 1)(x - 1)$

Problem 1.20 Factor $81x^4 - 16$.

(A) $(9x^2 + 4)(9x^2 - 4)$
(B) $(3x - 2)(3x - 2)(3x + 2)(3x + 2)$
(C) $(x - 2)(x + 2)(9x - 2)(9x + 2)$
(D) $(3x - 2)(3x + 2)(9x^2 + 4)$

1.3 Practice Questions

Problem 1.21 Factor the following:

(a) Factor $10x^2y^2 - 15xy^3 + 25xy^2z$.

(b) Factor $6x(a-b)^4 - 30x(b-a)^3$.

Problem 1.22 Factor $x^3 - 8y^3 - z^3 - 6xyz$.

Problem 1.23 Factor $a^{32} - b^{32}$.

Problem 1.24 Factor the following:

(a) Factor $2p^2 + p - 3$.

(b) Find all positive n such that $x^2 - nx - 12$ factors.

Problem 1.25 Factor $xy + xz + yw + wz$.

Problem 1.26 Factor $x^3 + x^2 + 4$

Problem 1.27 $a^3b - ab^3 + a^2 + b^2 + 1$.

Problem 1.28 $(m^2 - 1)(n^2 - 1) + 4mn$.

Problem 1.29 Factor $1 + 2x^2 + 2y^2 + 2x^2y^2 + x^4 + y^4$.

Problem 1.30 Factor $a^7 - a^5b^2 + a^2b^5 - b^7$.

2. Quadratics

Quadratic Equations

- A quadratic equation is an equation in x of the form $ax^2 + bx + c = 0$ where $a \neq 0$.
- If the coefficients a, b, c are given, the roots (solutions) are determined as well.
- On the other hand, if the roots are given, the coefficients a, b, c are also determined up to a common factor (that is to say, if a, b, c are multiplied by the same nonzero number, the roots don't change).
- This shows that in a quadratic equation, the roots and the coefficients have a very close relationship. This relationship is expressed in the discriminant and Vieta's Theorem.

Roots and the Discriminant

- **Roots of A Quadratic Equation**
  - There are several ways to solve a quadratic equation, such as factoring, completing the square, or the quadratic formula.
  - In general, the roots x_1 and x_2 are given by the quadratic formula:

$$x_{1,2} = \frac{-b \pm \sqrt{b^2 - 4ac}}{2a}.$$

  - You should know how to derive the quadratic formula.
- **Vertex of a Quadratic**
  - The maximum or minimum value of a quadratic function occurs at the *vertex*.

- The x-coordinate of the vertex is $\dfrac{-b}{2a}$. Note this is the average of the roots of the quadratic.

- **The Discriminant**
  - Note that the expression $b^2 - 4ac$ is inside the square root. Therefore the solution for the quadratic equation depends on the sign of this expression.
  - Denote $\Delta = b^2 - 4ac$. This is called the *discriminant*.
  - There are three possibilities:
    1. If $\Delta > 0$, the equation has two distinct real roots.
    2. If $\Delta = 0$, the equation has exactly one real root, or we say it has two identical roots, or double roots.
    3. If $\Delta < 0$, the equation has no real roots.

Vieta's Theorem

- Vieta's Theorem (or Vieta's Formulas) shows the relation between the roots and the coefficients.
- If x_1 and x_2 are the roots for the quadratic equation $ax^2 + bx + c = 0 (a \neq 0)$, then

$$x_1 + x_2 = -\frac{b}{a}, \quad \text{and} \quad x_1 x_2 = \frac{c}{a}.$$

- These relations can be easily proved using the quadratic formula.
- In the AMC 10 and 12, many problems can be easily solved using discriminant or Vieta's Theorem.

2.1 Example Questions

Problem 2.1 Determine the number of roots without solving the equation:

(a) $3x^2 + 4x - 5 = 0$

(b) $4x^2 + 20x + 25 = 0$

(c) $2x^2 + 2x + 3 = 0$

Problem 2.2 For each of the following, state whether the function has a maximum or a minimum and find the maximum or minimum value.

(a) $f(x) = 4x^2 - 12x - 16$.

(b) $f(x) = 61 + 72x - 36x^2$.

(c) $f(x) = x^4 - 4x^2 + 3$.

Problem 2.3 For what x are the following functions positive?

(a) $f(x) = 4x^2 - 12x - 16$.

(b) $f(x) = -x^2 + 4x + 2$.

(c) $f(x) = 2x^2 + 2x + 3$

(d) $f(x) = x^4 - 4x^2 + 3$.

Problem 2.4 Given that the equation $x^2 - 2x - m = 0$ has no real roots, how many real roots does the equation $x^2 + 2mx + 1 + 2(m^2 - 1)(x^2 + 1) = 0$ have?

Problem 2.5 For what values of m does the equation $4x^2 + 8x + m = 0$ have two distinct real roots?

Problem 2.6 Use Vieta's formula to solve the following.

(a) A quadratic equation has two roots $\dfrac{2}{3}$ and $-\dfrac{1}{2}$, what is this equation? (multiple answers are possible)

(b) Two real numbers have sum -10 and product -5, find these two numbers.

Problem 2.7 Answer the following.

(a) (2008 AMC 10B) A quadratic equation $ax^2 - 2ax + b = 0$ has two real solutions. What is the average of the solutions?

(b) (2005 AMC 10B/12B) The quadratic equation $x^2 + mx + n = 0$ has roots that are twice those of $x^2 + px + m = 0$, and none of m, n, and p is zero. What is the value of n/p?

Problem 2.8 Let x_1, x_2 be the two roots for equation $x^2 + x - 3 = 0$, find the value of $x_1^3 - 4x_2^2 + 19$.

Problem 2.9 Given the equation in x, $x^2 + 2mx + m + 2 = 0$:

(a) For what values of m does the equation have two (not necessarily distinct) positive roots?

(b) For what values of m does the equation have one positive root and one negative root?

Problem 2.10 (2005 AMC 10A/12A) There are two values of a for which the equation $4x^2 + ax + 8x + 9 = 0$ has only one solution for x. What is the sum of those values of a?

2.2   Quick Response Questions

Problem 2.11 Factor $x^2 + 6x + 8$. What is the smallest root of $x^2 + 6x + 8 = 0$.

Problem 2.12 Factor $x^2 + (m+2)x + 2m$. There is an integer L such that L is a root of $x^2 + (m+2)x + 2m = 0$ for all m. What is L?

Problem 2.13 What is the discriminant of the quadratic $3x^2 + 3x + 4$?

Problem 2.14 The roots of $x^2 - 3x + 1 = 0$ are of the form $\dfrac{A \pm \sqrt{B}}{2}$. What is B?

Problem 2.15 For $m = \pm\sqrt{K}$, $2x^2 + mx + 3 = 0$ has exactly one real root. What is K?

Problem 2.16 What is the x-coordinate of the vertex of the quadratic $4x^2 + 24x + 48$?

Problem 2.17 The vertex of the quadratic $x(x - 2a)$ occurs when $x = 4$. The vertex of the quadratic $(x + a)(x - 3a)$ occurs when $x = L$. What is L?

Problem 2.18 For how many integers x is $f(x) = x^2 + x - 6$ negative?

Problem 2.19 What is the sum of the roots of $49x^2 - 196x + 193$?

Problem 2.20 The quadratic $2x^2 - Bx + C = 0$ has roots 2 and $1/2$. What is C?

2.3 Practice Questions

Problem 2.21 Determine the number of roots without solving the equation:

(a) $3x^2 - 4x + 5 = 0$

(b) $x^2 + 2x - 15 = 0$

(c) $9x^2 - 24x + 16 = 0$

Problem 2.22 A batter hits a baseball. The height of the baseball in feet after t seconds is $h = 3 + 64t - 16t^2$.

(a) When does the baseball reach its maximum height?

(b) What is the maximum height of the baseball?

Problem 2.23 For what values of x is $f(x) = 2x^2 - 7x - 15$ non-negative?

Problem 2.24 For what values of m does $x^2 - 4x - m = 0$ have no real solutions while $x^2 - 9x + m^2 = 0$ has at least one real solution?

Problem 2.25 For what values of m does $x^2 + mx + m = 0$ have a real solution?

Problem 2.26 A rectangle has perimeter 24 and area 35. Find the side lengths of the rectangle.

Problem 2.27 Suppose that $ax^2 + bx + c = 0$ has real solutions x_1 and x_2. Find the solutions to the equation $a(x + x_1)^2 + b(x + x_1) + c = 0$.

Problem 2.28 Let x_1 and x_2 be the two roots of $17x^2 - 8x - 2 = 0$. Use Vieta's Theorem to find $x_1^2 + x_2^2$.

Problem 2.29 Find all m such that $x^2 - (2m - 3)x + m(m - 3) = 0$ has one positive and one negative root.

Problem 2.30 Find all a such that $(a + 1)x^2 + (a - 1)x - 2 = 0$ has exactly one real solution.

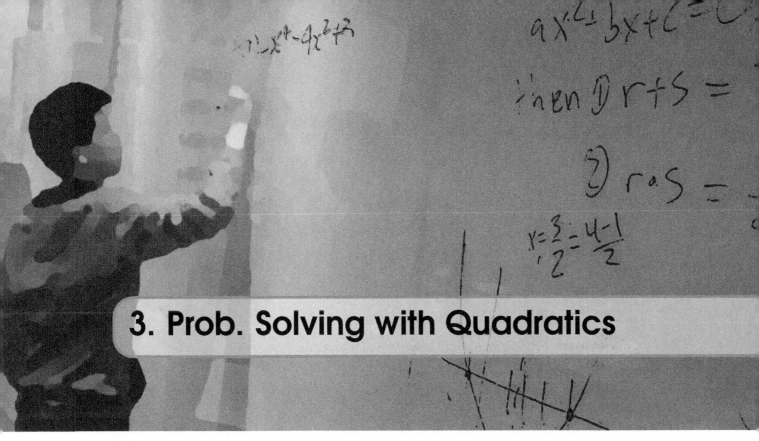

3. Prob. Solving with Quadratics

Review of Quadratics of the form $ax^2 + bx + c$

- The discriminant $\Delta = b^2 - 4ac$. There are three possibilities:
    - i) If $\Delta > 0$, the equation has two distinct real roots;
    - ii) If $\Delta = 0$, the equation has exactly one real root; or we say it has two identical roots, or double roots.
    - iii) If $\Delta < 0$, the equation has no real roots.
- Vieta's Theorem: The sum of the two roots of a quadratic is $-b/a$ and the product of the two roots is c/a.

3.1 Example Questions

Problem 3.1 Answer the following.

(a) How many real roots does the equation

$$(7x - 13)(x^2 + 7x - 13)(x^2 - 7x + 13)$$

have?

(b) Given $p > 0$ and $q < 0$, how many positive roots does the equation $x^2 + px + q = 0$ have?

(c) Without solving the equation, find the number of real roots for x: $(n^2 + 1)x^2 - 2nx + (n^2 + 4) = 0$.

Problem 3.2 Find k in each of the following scenarios:

(a) In the equation $x^2 - 402x + k = 0$, one of the roots plus three equals 80 times the other root.

(b) Let x_1 and x_2 be the two roots of the equation $4x^2 - 8x + k = 0$. Suppose further that $\frac{1}{x_1} + \frac{1}{x_2} = \frac{8}{3}$.

Problem 3.3 Do the following.

(a) The sum of squares of the roots of equation $x^2 + 2kx = 3$ is 10. Find the possible values of k.

(b) For equation $2x^2 + mx - 2m + 1 = 0$, the sum of squares of the two real roots is $\frac{29}{4}$. Find the value of m.

Problem 3.4 The quadratic equation $x^2 + 2kx + 2k^2 - 1 = 0$ has at least one negative root. Find the possible range of values for k.

Problem 3.5 The two real roots of $x^2 + (m-2)x + 5 - m = 0$ are both greater than 2. Find the possible range of values for real number m.

Problem 3.6 Find all ordered pairs (a, b) such that $a^2 + b^2$ is prime, and the equation $x^2 + ax + 1 = b$ has two positive integer roots.

Problem 3.7 The equation $x^2 + (a-6)x + a = 0$ has two integer roots. Find the value of a.

Problem 3.8 More Vieta Practice.

(a) If x_1 and x_2 are the two real roots of $x^2 + x + q = 0$, and $|x_1 - x_2| = q$, find the value of q.

(b) For the equation $x^2 + mx + n = 0$, the difference between the two roots is p and the product of the two roots is q. What is $m^2 + n^2$ in terms of p and q?

Problem 3.9 Find all real solutions to the system of equations: $x + y = 2$ and $xy - z^2 = 1$. Justify your answer.

Problem 3.10 If x_1 and x_2 are integer roots of the equation $x^2 + mx + 2 - n = 0$, and $(x_1^2 + 1)(x_2^2 + 1) = 10$, how many possible pairs (m, n) are there?

3.2 Quick Response Questions

Problem 3.11 A rectangular room is 8 feet longer than it is wide and has an area of 240 square feet. Find the length of the room in feet.

Problem 3.12 $x^2 - 22x + 21 = (x - 11)^2 + K$ for an integer K. What is K? (Recall this is part of the process of completing the square.)

Problem 3.13 Consider the equations $ax^2 + bx + c = 0$ and $y^2 + by + ac = 0$, where $a \neq 0$. Is there a relation between their roots x_1, x_2 and y_1, y_2?

(A) No, the roots are not related.
(B) Yes, $y_1 = x_1, y_2 = x_2$.
(C) Yes, $y_1 = ax_1, y_2 = ax_2$
(D) Yes, $y_1 = x_1 + a, y_2 = x_2 + a$.

Problem 3.14 Find a so that $ax^2 + 30x + 25 = 0$ has a double root. What is a?

Problem 3.15 Let Δ_1 be the discriminant of $a(x+2)^2 + b(x+2) + c = 0$. How does it compare to Δ_2, the discriminant of $ax^2 + bx + c = 0$?

(A) $\Delta_1 = \Delta_2$.
(B) $\Delta_1 = 2 \times \Delta_2$
(C) $2 \times \Delta_1 = \Delta_2$
(D) $\Delta_1 = \Delta_2 + 2$.

Problem 3.16 Find two numbers whose sum is 36 and whose product is 323. What is the smaller number?

Problem 3.17 Let p, q be roots of $x^2 + 4x - 7 = 0$. What is $p^2 + q^2$?

Problem 3.18 Let r, s be roots of $2x^2 - 4x + 1 = 0$. What is $\dfrac{1}{r} + \dfrac{1}{s}$?

Problem 3.19 Let $a > b$ be the two roots of $2x^2 - 4x - 16 = 0$. What is $a^2 - 2b$?

Problem 3.20 Let m, n, o, p be the roots to the equation $(x^2 + 4x + 2)^2 = 0$. What is $mnop$?

3.3 Practice Questions

Problem 3.21 Without solving the equation, find the number all integers n such that the equation $(n+1)x^2 - 2nx + (n+4) = 0$. has a real root for x.

Problem 3.22 Find a quadratic equation whose roots satisfy $x_1 + x_2 = 4$, $\dfrac{1}{x_1} + \dfrac{1}{x_2} = 2$ and solve for x_1, x_2.

Problem 3.23 Suppose that a and b are real numbers and that $x^2 + ax + b = 0$ has two real solutions x_1 and x_2 with $x_1^2 + x_2^2 = 1$. Find all possible values of b and express a in terms of b.

Problem 3.24 Find all k such that $x + 2\sqrt{x} + k = 0$ has a real solution and then solve for x.

Problem 3.25 Find all m such that the difference of the roots of $2x^2 - mx - 8 = 0$ is $m - 1$.

Problem 3.26 Find all ordered pairs (a, b) such that $a^2 + b^2 = 34$, and the equation $x^2 + ax + 1 = b$ has two positive integer roots.

Problem 3.27 The equation $x^2 + (a + 14)x - a = 0$ has two integer roots. Find the possible values for a.

Problem 3.28 Derive the quadratic formula from Vieta's formulas.

Problem 3.29

(a) Find all values of m for which the system of equations $y = mx + \sqrt{2}$ and $x^2 + y^2 = 1$ has real solutions. For what values of m is there exactly one solution?

(b) Find (x, y) in Part (a) for those m for which the system has exactly one solution.

Problem 3.30 Suppose x_1 and x_2 are real integer roots of the equation $x^2 - mx + n = 0$ and $(1 + x_1)(1 + x_2) = mn$. Find all possible pairs (m, n) and the corresponding solutions $x_{1,2}$.

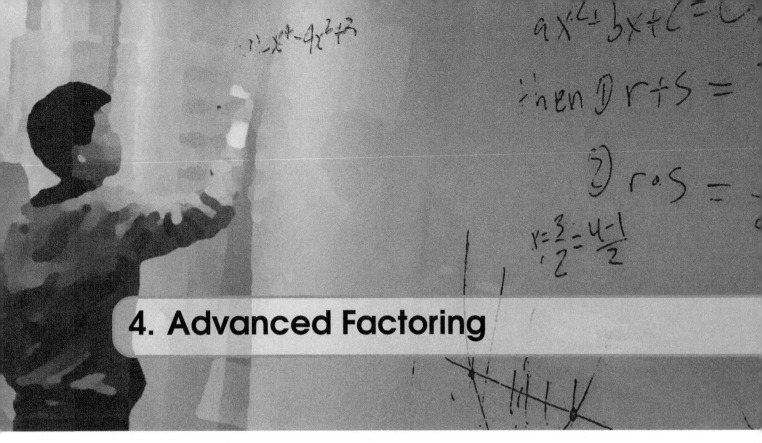

Change of Variables

- A substitution or change of variables is often useful to replace a complicated part of the original expression, to make a simpler new expression.
- As a simple example, we can factor $x^4 + 2x^2 - 3$ as $(x^2 - 1)(x^2 + 3) = (x - 1)(x + 1)(x^2 + 3)$ using the substitution $x^2 = z$, so the equation becomes $z^2 + 2x - 3$ and can be factored as $(z - 1)(z + 3)$.

4.1 Example Questions

Problem 4.1 Factor the following.

(a) $(x^2 + x + 1)(x^2 + x + 2) - 12$. Hint: Try letting $y = x^2 + x + 1$.

(b) $(x^2 + 3x + 2)(x^2 + 7x + 12) - 120$. Hint: Factor and regroup so you can make the substitution $x^2 + 5x + 5$.

Problem 4.2 Factor the following using a change of variables.

(a) $x^2 + x - 14 - \dfrac{1}{x} + \dfrac{1}{x^2}$. Hint: Note $\left(x - \dfrac{1}{x}\right)^2 = x^2 - 2 + \dfrac{1}{x^2}$.

(b) $6x^4 + 7x^3 - 36x^2 - 7x + 6$.

Problem 4.3 Factor $(x+3)(x^2-1)(x+5) - 20$

Problem 4.4 Factor $(x^2 + xy + y^2)^2 - 4xy(x^2 + y^2)$. Hint: Let $u = x + y, v = xy$.

Problem 4.5 Factor $x^3 + 3x^2 - 4$

Problem 4.6 Factor $(x^2 + 4x + 8)^2 + 3x(x^2 + 4x + 8) + 2x^2$.

Problem 4.7 Factor $a^2 + (a+1)^2 + (a^2 + a)^2$

Problem 4.8 Factor the following.

(a) $2acx + 4bcx + adx + 2bdx + 4acy + 8bcy + 2ady + 4bdy$

(b) $1 + 2a + 3a^2 + 4a^3 + 5a^4 + 6a^5 + 5a^6 + 4a^7 + 3a^8 + 2a^9 + a^{10}$.

Problem 4.9 Factor $a^5 + a + 1$.

Problem 4.10 Evaluate the following: $\dfrac{(1994^2 - 2000)(1994^2 + 3985) \times 1995}{1991 \cdot 1993 \cdot 1995 \cdot 1997}$.

4.2 Quick Response Questions

Problem 4.11 Which of the following substitutions for z gives $(x+1)(x+3)+4 = z^2 + 2z + 4$?

(A) $z = x - 1$
(B) $z = x + 1$
(C) $z = x + 3$
(D) $z = x - 3$

Problem 4.12 $8x^3 - 36x^2y + 54xy^2 - 27y^3$ can be factored as $(Sx + Ty)^3$. What is $S + T$?

Problem 4.13 The equation $x^6 - 3x^3 - 40 = 0$ has one (real) integer root. What is this root?

Problem 4.14 $x^4 - 10x^2 + 24$ can be fully factored as $(x - a)(x - b)(x^2 - c)$. What is c?

Problem 4.15 Find the roots of $\dfrac{1}{x^2} - \dfrac{4}{x} + 3 = 0$. What is the sum of the roots? Round your answer to the nearest hundredth if necessary.

Problem 4.16 Factor $1 + a + a^2 + a^3$.

(A) $(a+1)(a^2+a+1)$
(B) $(a-1)(a^2+1)$
(C) $(a+1)(a^2+1)$
(D) $(a+1)(a-1)(a+1)$

Problem 4.17 Factor $(x^2+3x+1)(x^2+3x+4)+2$.

(A) $(x+1)(x+2)(x^2+3x+3)$
(B) $(x+1)(x+2)(x^2+3x+1)$
(C) $(x+1)(x+2)(x+3)(x+4)$
(D) $(x^2+3x-1)(x^2+3x+3)$

Problem 4.18 Factor $8 - \dfrac{36}{x} + \dfrac{54}{x^2} - \dfrac{27}{x^3}$.

(A) $(2/x-3)^3$
(B) $(3-2/x)^3$
(C) $(3/x-2)^3$
(D) $(2-3/x)^3$

Problem 4.19 Which of the following is equal to $x^3y + xy^3$ after the substitution $u = x+y$ and $v = xy$.

(A) $uv(u-2v)$
(B) $v(u^2-2v)$
(C) u^2+v
(D) uv^2

Problem 4.20 Write $\dfrac{2017^2 + 2017 + 1}{2017^3 - 1}$ as the reduced fraction $\dfrac{P}{Q}$ (with $P, Q \geq 1$ and $\gcd(P, Q) = 1$). What is $P + Q$?

4.3   Practice Questions

Problem 4.21 Factor $(x^2 + 3x + 2)(4x^2 + 8x + 3) - 90$.

Problem 4.22 Factor $x^4 + 7x^3 + 14x^2 + 7x + 1$

Problem 4.23 Factor $(x+1)(x+3)(x+5)(x+7) + 15$

Problem 4.24 Factor $(x+y)^4 - 6xy(x^2 + y^2) - 4x^2 y^2$.

Problem 4.25 Factor $x^3 + 9x^2 + 26x + 24$

Problem 4.26 Factor $(2x^2 - 3x + 1)^2 - 22x^2 + 33x - 1$

Problem 4.27 Factor $ab(c^2 - d^2) - cd(a^2 - b^2)$

Problem 4.28 Factor $abc + 3b^2 c + 3abd + 9b^2 d + 3ac^2 + 9bc^2 + 9acd + 27bcd$.

Problem 4.29 Factor $a^5 + a^4 + 1$.

Problem 4.30 Evaluate

$$\frac{(2016^2 - 4032)(2016^2 + 2014)}{2014 \times 2016 \times 2018}.$$

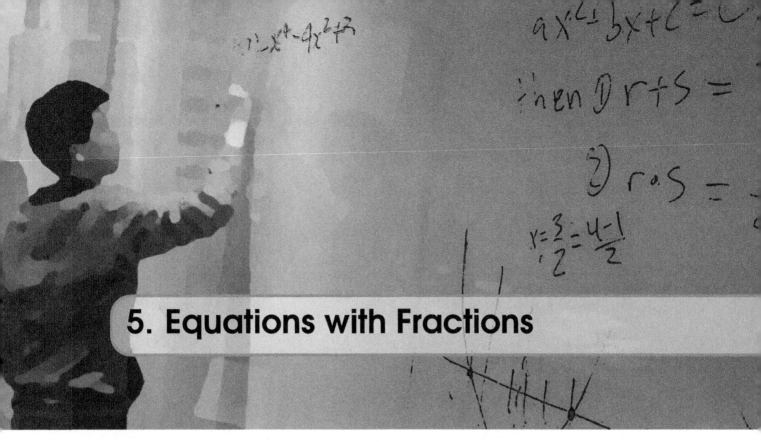

5. Equations with Fractions

Solving Equations

- The standard method to solve equations is by factoring.
- Methods from previous weeks, such as change of variables are very helpful for factoring equations which can then lead to solutions.
- The goal of most substitutions is to turn the equation into a quadratic, which we can then solve by factoring or by using the quadratic formula.

Equations with Fractions

- When unknowns appear on denominators, the standard method is to multiply everything by the denominator (or the least common multiple of the denominators) to get rid of the denominators.
- This is likely to introduce extraneous roots. Therefore, it is always necessary to verify the roots at the end.
- As with other types of equations, a change of variables may be used to simplify the equations.

5.1 Example Questions

Problem 5.1 Find the real solutions to $(2x^2 - 3x + 1)^2 = 22x^2 - 33x + 1$.

Problem 5.2 Solve $(x^2 + x - 1)^2 + 2x(x^2 + x - 1) + x^2 = 4$.

Problem 5.3 Solve $3x^2 + (x + 2)^2 + (x^2 + x)^2 = 9$ over the reals.

Problem 5.4 Solve the following:

(a) $\dfrac{15}{x+1} = \dfrac{15}{x} - \dfrac{1}{2}$.

(b) $\dfrac{4x}{x^2 - 4} - \dfrac{2}{x - 2} = \dfrac{x+1}{x+2}$.

Problem 5.5 Solve the following:

(a) $\dfrac{3 - x}{2 + x} = 5 - \dfrac{4(2 + x)}{3 - x}$.

(b) $\dfrac{x - 3}{x + 1} - \dfrac{x + 1}{3 - x} = \dfrac{5}{2}$.

Problem 5.6 Solve the equation $\dfrac{1}{2x^2 - 3} - 8x^2 + 12 = 0$.

Problem 5.7 Solve: $\left(\dfrac{x+1}{x^2 - 1}\right)^2 - 4\left(\dfrac{x+1}{x^2 - 1}\right) + 3 = 0$.

Problem 5.8 Solve $2x^4 - 9x^3 + 14x^2 - 9x + 2 = 0$.

Problem 5.9 Solve $(x^2 - 1)(x^2 + 10x + 24) = 24$.

Problem 5.10 Solve the equation $\dfrac{x-1}{x+1} + \dfrac{x-4}{x+4} = \dfrac{x-2}{x+2} + \dfrac{x-3}{x+3}$.

5.2   Quick Response Questions

Problem 5.11 What is the result after making the substitution of $z = x^2 + 3x + 2$ in the equation

$$\frac{1}{x^2 + 3x + 2} - 2x^2 - 6x = 4?$$

(A) $z^{-1} + 2z = 8$

(B) $z^{-1} - 2z = 0$

(C) $z^{-1} + 2z = 0$

(D) $z^{-1} - 2z = 8$

Problem 5.12 Solve for x if $\dfrac{1}{x^2 + x - 2} + \dfrac{1}{x^2 - 3x - 10} = 0.$

Problem 5.13 Find the sum of all solutions to $\dfrac{x+5}{x-5} = \dfrac{x+6}{x-6}.$

Problem 5.14 Find the product of all solutions to $\left(\dfrac{1}{y-1}\right)^2 + \dfrac{1}{y-1} = 6.$ Round your answer to the nearest hundredth if necessary.

Problem 5.15 The equation $x^3 + \dfrac{1}{x^3} + 3x + \dfrac{3}{x} = 8$ has exactly one solution. What is it? Round your answer to the nearest integer if necessary.

Problem 5.16 What is the product of all the real solutions to $(x^2 + 2)(x^2 + 3) - 12 = 0?$

Problem 5.17 If $z^2 - x^2z + xz - x^3 = 0$ then

(A) $z = x^2$ or $z = x$
(B) $z = x^2$ or $z = -x$
(C) $z = -x^2$ or $z = x$
(D) $z = -x^2$ or $z = -x$

Problem 5.18 Find the smallest positive real root of $\dfrac{1}{x^2} - \dfrac{4}{x} + 4 = 0$. Round your answer to the nearest tenth if necessary.

Problem 5.19 Solve $\dfrac{1}{x} + \dfrac{1}{x+2} = \dfrac{12}{5}$. If $r < s$ are the roots, what is $r - s$ rounded to the nearest tenth?

Problem 5.20 Solve $\dfrac{1}{x^2} - \dfrac{1}{x^2+3} = \dfrac{3}{4}$. What is the difference between the largest and smallest root?

5.3 Practice Questions

Problem 5.21 Solve $4(2x^2 - 1)^2 + 8x^2 - 3 = 0$.

Problem 5.22 Solve $4(x+1)^4 + 3x(x+1)^2 = x^2$.

Problem 5.23 Solve $x^3 + 6x^2 + 13x + 10 = 0$ over the reals.

Problem 5.24 Solve (over the reals) $\dfrac{1}{x-1} + \dfrac{1}{x+1} = 3$.

Problem 5.25 Solve: $\dfrac{3x-1}{x^2+1} - \dfrac{3x^2+3}{3x-1} = 2$.

Problem 5.26 Find the real solutions to $\dfrac{1}{x^2+2} - x^2 = \dfrac{1}{2}$.

Problem 5.27 Find all real solutions to $\dfrac{x+2}{x^2-4} + \dfrac{x+1}{x^2-x-2} = \dfrac{x}{x-2}$.

Problem 5.28 Find all real roots of $x^4 - 2x^3 - x^2 - 2x + 1 = 0$.

Problem 5.29 Find the real roots of $(x^2 + x + 1)(x^2 + x + 2) = 12$.

Problem 5.30 Solve $\dfrac{x+3}{x-3} - \dfrac{x+2}{x-2} = 1$.

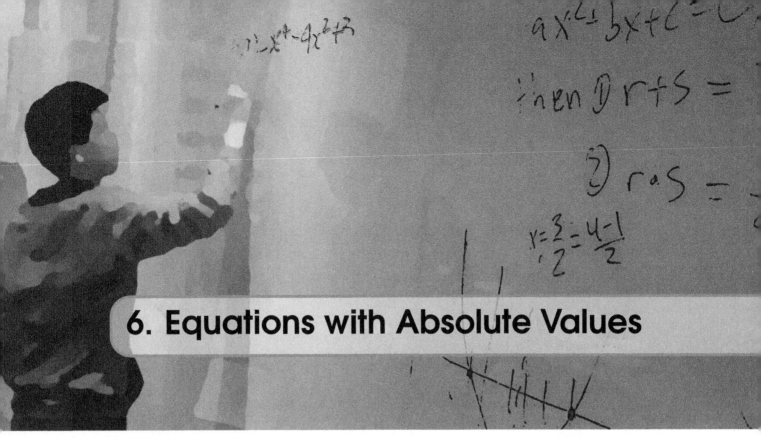

Equations with Absolute Values

- The absolute value of a real number a is defined as:

$$|a| = \begin{cases} a, & \text{if } a \geq 0; \\ -a, & \text{if } a < 0. \end{cases}$$

- The standard method to deal with absolute values is case analysis: solve in intervals where the expressions inside the absolute value do not change signs.
- Sometimes the following techniques also help:
  - Change of variables.
  - Using the property that absolute values are always nonnegative.

6.1 Example Questions

Problem 6.1 Graph the following equations. Where do they cross the x-axis?

(a) $y = \left| \dfrac{x}{2} - 2 \right| - 2.$

(b) $y = |2x^2 + 5x - 12|$

Problem 6.2 Solve the following equations over the reals by considering cases.

(a) $|x| + 2 = |2x|$.

(b) $|x^2 + 1| = 2|x - 1|$.

Problem 6.3 Solve the equation $|x| - 2 = -|1 - x|$.

Problem 6.4 Solve $|x - |2x + 1|| = 3$.

Problem 6.5 Solve the following

(a) Solve the equation $|x^2 - 11x + 10| = |2x^2 + x - 45|$.

(b) $\dfrac{|x+4|}{|x+1|} = \dfrac{|x+3|}{|x+2|}$.

Problem 6.6 If $|m - 2009| = -(n - 2010)^2$, what is $(m - n)^{2011}$?

Problem 6.7 Solve the equation $|x^2 + 6x + 1| = |(x + 3)^2 - 4|$.

Problem 6.8 The equation $|x^2 - 5x| = a$ has exactly two distinct real roots. What is the possible range of values for a?

Problem 6.9 Find all solutions to $|||x+1| - 1| - 1| = 1$.

Problem 6.10 Consider the function $f(x) = |x^2 - 2x - 3| - |x^2 + x - 2|$.

(a) Break $f(x)$ into cases to write an equation for $f(x)$ without absolute values.

(b) Find the domain, range, and any zeros of $y = f(x)$.

6.2   Quick Response Questions

Problem 6.11 What expression is the same as $4|x+3|$?

   (A) $4x+12$
   (B) $|4x+3|$
   (C) $|4x+12|$
   (D) $4x+3$

Problem 6.12 What expression is not the same as $|6x-8y|$?

   (A) $|8y-6x|$
   (B) $2|3x-4y|$
   (C) $6|x|-8|y|$
   (D) $|-2||3x-4y|$

Problem 6.13 The minimum value of $y=|4x^2-12x-7|$ occurs at $x=-\dfrac{A}{2}$ and at $x=\dfrac{B}{2}$ where A and B are positive integers. What is $A+B$?

Problem 6.14 What is the minimum value of $y=|x^2+4x+7|$?

Problem 6.15 Solve $|x+3|=|x|$. What is the smallest solution, rounded to the nearest hundredth if necessary?

Problem 6.16 How many integer solutions are there to the inequality $|3x|\le 9$?

Problem 6.17 Solve $|x^2 + 2| = 6$ over the real numbers. The solutions are $\pm K$ for an integer K. What is K?

Problem 6.18 Solve $|x + |x + 1|| = 4$. How many solutions are there?

Problem 6.19 Find all triples (x, y, z) such that $|x - 2| + |y^2 - 4| + |z - x| = 0$. For these triples, what is the largest possible value of $x + y + z$?

Problem 6.20 What is the smallest positive integer that is a solution to the inequality $|3x - 4| > 12$

6.3   Practice Questions

Problem 6.21 Graph the equation $y = 4 - 2|x + 1|$. Where does it cross the x-axis?

Problem 6.22 $|x + 2| + |3(x + 2)| = 8$

Problem 6.23 Find all the real solutions to $|x^2 - 1| = |x + 1|$.

Problem 6.24 $|3 + |x + 2|| = 5$

Problem 6.25 Find the real solutions of $|x^2 - 3x + 1| = |x^2 + x + 2|$.

Problem 6.26 Solve the equation $|x^2 - 1| + |x^2 + x - 2| = 0$.

Problem 6.27 Solve the equation $|y^2 + y| = 2y^2 + 2y - 2$

Problem 6.28 For what values of a, b the equation $x^2 + 2(1 + a)x + (3a^2 + 4ab + 4b^2 + 2) = 0$ has real roots?

Problem 6.29 Solve for x: $||x + 3| - 3| = 3$

Problem 6.30 Find the domain, range, and any zeros of $y = ||x+2|-3|-1$.

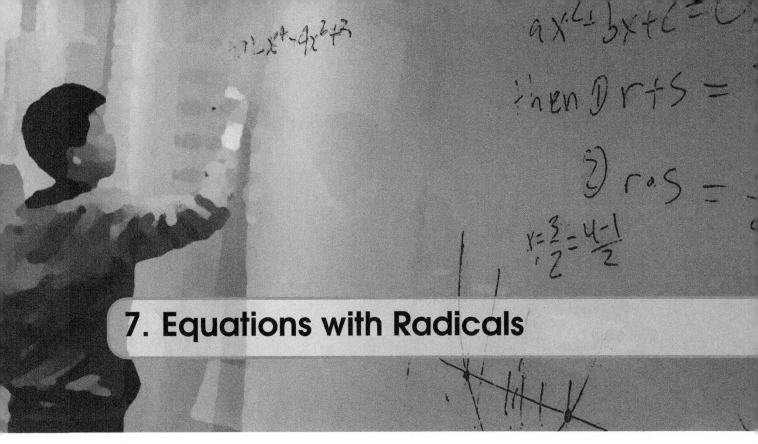

7. Equations with Radicals

Equations with Radicals

- If unknown variables appear inside radicals, the common method is to square (or cube, etc., depending on the order of the roots) both sides to remove the radicals.
- Sometimes the following methods also help:
  - Change of variables
  - Using the fact that $\sqrt{a}$ is always nonnegative for $a \geq 0$.

7.1   Example Questions

Problem 7.1 Find the domain and range of the following functions.

(a) $y = \sqrt{x^2 + 3x - 4}$.

(b) Find the domain and range of $y = \sqrt{x^2 - 6x + 13}$.

Problem 7.2 Find the real solutions to the following.

(a) $3 - \sqrt{2x - 3} = x$.

(b) $\sqrt{x + 3} - \sqrt{3x - 2} = -1$.

Problem 7.3 Solve: $\sqrt{x^2 + 3x + 7} - \sqrt{x^2 + 3x - 9} = 2$.

Problem 7.4 Solve: $x^2 - \sqrt{3x^2 + 7} = 1$.

Problem 7.5 Solve: $\sqrt{\sqrt{x + 4} + 4} = x$

Problem 7.6 Solve for real x: $\sqrt{\dfrac{x - 2}{x + 2}} + \sqrt{\dfrac{9x + 18}{x - 2}} = 4$.

Problem 7.7 Solve for x: $(x - \sqrt{3})x(x + 1) + 3 - \sqrt{3} = 0$.

Problem 7.8 Let a be a real number, and the equation $x^2 + a^2x + a = 0$ has real roots for x. Find the maximum possible root x.

Problem 7.9 Solve $\sqrt{2x + 2} - \sqrt{x + 3} = \sqrt{x + 1} - \sqrt{2x + 4}$.

Problem 7.10 For what range of k does $\sqrt{2x^2 + 4} = x + k$ have real solutions?

7.2 Quick Response Questions

Problem 7.11 Solve $\sqrt{x^2 - 16} + |x - 4| = 0$. What is the largest solution?

Problem 7.12 Consider the equation $|x - 4| = \sqrt{x}$. Which of the following quadratics has the same solutions?

(A) $x^2 - 8x + 16 = 0$
(B) $x^2 - 7x + 16 = 0$
(C) $x^2 - 9x + 16 = 0$
(D) $x^2 + 9x - 16 = 0$

Problem 7.13 The domain of $\sqrt{x + 5} + \sqrt{7 - x}$ can be written as the closed interval $[A, B]$. What is $A + B$?

Problem 7.14 The range of $y = \sqrt{x^2 - 16} + |x|$ is all y such that $y \geq K$ for an integer K. What is K?

Problem 7.15 How many solutions does $\sqrt{1 + \sqrt{1 + x}} = 2$ have?

Problem 7.16 The equation $\sqrt{|x| + x} = 8$ has one solution. What is it? Round your answer to the nearest tenth if necessary.

Problem 7.17 The equation $\sqrt{x + 5} + \sqrt{x + 7} = 5$ has one solution of the form $\dfrac{P}{Q}$ for $P > 1$ and $Q > 1$ with $\gcd(P, Q) = 1$. What is $P - Q$?

Problem 7.18 The equation $\dfrac{1}{\sqrt{x+2}} + \sqrt{x+2} = \dfrac{10}{3}$ has one solution of the form $\dfrac{P}{Q}$ for P an integer and $Q > 1$ with $\gcd(P,Q) = 1$. What is $P - Q$?

Problem 7.19 Solve the following equation for y, $-x^3 - x^2y - 2x^2 + xy - 2x + y^2 - 4 = 0$.

   (A) $y = x^2 + 2$ or $y = -x - 2$
   (B) $y = x^2 - 2$ or $y = x - 2$
   (C) $y = x^2 + x + 2$ or $y = -x - 2$
   (D) $y = -x^2 - x + 2$ or $y = x + 2$

Problem 7.20 Find the smallest real solution of $-x^3 - x^2 - 2x^2 + x - 2x + 1 - 4 = 0$.

7.3 Practice Questions

Problem 7.21 Find the domain and range of $y = \sqrt{81 - 9x^2} - \sqrt{9 - x^2}$.

Problem 7.22 Solve $\sqrt{x - 4} = x - 6$ for real values of x.

Problem 7.23 Solve $\sqrt{x^2 + 2x + 6} = \sqrt{x^2 + 6x + 3}$

Problem 7.24 Solve: $2x^2 - \sqrt{4x^2 - 12x} = 6x + 4$.

Problem 7.25 Solve for x: $\sqrt{5 - \sqrt{5 - x}} = x$.

Problem 7.26 Solve the equation $\sqrt{\dfrac{x}{x + 2}} + \sqrt{\dfrac{x + 2}{x}} = 2$.

Problem 7.27 Solve $16 + 4x - 4x^2 - x^3 = 0$.

Problem 7.28 Recall Problem 7.8. For what value of a is the maximum root achieved?

Problem 7.29 Find the number of solutions to $\sqrt{x + 2} - \sqrt{x + 3} = \sqrt{x + 1} - \sqrt{x + 4}$.

Problem 7.30 The equation $2kx^2 + (8k+1)x + 8k = 0$ has two distinct real roots for x. Find the range of values for k.

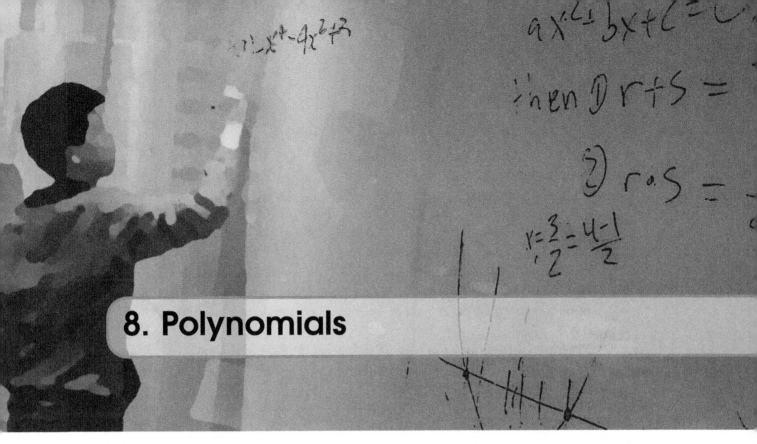

8. Polynomials

General Definitions

- **Polynomial.** A function that is made of adding multiples of powers of a variable. Examples: $5x^3 - 2x + 1$, $-a^{99} - 9a^9 - 99$, and $4z^2 - 12z + 9$. Usually polynomials are written with the powers going up or going down.
- **Monomial (or term).** Both words refer to a polynomial with exactly one piece. Ever polynomial can be thought of as a sum of monomials / terms. For example, the polynomial $x^7 - 4x^5 + 18x$ is the sum of the monomials x^7, $-4x^5$, and $18x$.
- **Degree (of a polynomial).** The largest power in a polynomial. The degrees of the polynomials $5x^3 - 2x + 1$, $-a^{99} - 9a^9 - 99$, and $4z^2 - 12z + 9$ are 3, 99, and 2. A constant is also a polynomial, and it has degree 0 if the constant itself is nonzero.
- **The zero polynomial.** The function $p(x) = \mathbf{0}$ is a polynomial as well, called the "zero polynomial". Note that this is not an equation. The polynomial always has value 0 no matter what x is. The zero polynomial has no terms and, strictly speaking, it has no degree either. Sometimes it is convenient to define the degree of the zero polynomial to be either -1 or negative infinity ($-\infty$).
- **Coefficient (of a term).** The number being multiplied by a power. The coefficient of x^3 in $5x^3 - 2x + 1$ is 5; the coefficient of a^9 in $-a^{99} - 9a^9 - 99$ is -9; the coefficient of z^7 in $4z^2 - 12z + 9$ is 0.
- **Function notation.** Often it is convenient to use function notation to represent long polynomials. When we say $f(x) = 5x^3 - 2x + 1$, we are saying that f "assigns" to a number x the value $5x^3 - 2x + 1$.
- **Zero of a polynomial** (also called **"root"** of a polynomial equation). The value

of x where the polynomial $P(x)$ has the value 0. It is a solution of the polynomial equation $P(x) = 0$.

- **Factored polynomial.** A polynomial is sometimes written as a product of different polynomials. For example, $(3x + 1)(2x - 7)$, or $(z^2 + 2z + 3)(-2z + 5)(7z^7 + 4z^4)$, or $(2y - 8)^2(5y^2)^3$. This is sometimes done to make equations easier to solve, or to make values easier to compute.

- **Polynomial of multiple variables.** There can be multiple variables in a polynomial: $3xy^2$, $9x^5y^5 + 20x^3y^4$. The **degree** of these polynomials is the maximum sum of the exponents of variables. For example, $3xy^2$ has degree 3, and $9x^5y^5 + 20x^3y^4$ has degree 10.

- **Polynomial long division**: Given polynomials $P(x)$ and $D(x)$, we can always write $P(x) = D(x)Q(x) + R(x)$, where $\deg R(x) < \deg D(x)$. Here $Q(x)$ is the quotient, and $R(x)$ is the remainder.

Important Theorems

- **Fundamental Theorem of Algebra** (also known as the **Gauss-d'Alembert Theorem**.
  - Every nonconstant polynomial with complex coefficients has at least one complex zero.
  - Consequently, the number of zeros of a polynomial equals the degree, multiplicities counted.
  - Thus, a polynomial of degree n has at most n real zeros.
  - Note: The proof of this theorem is beyond the scope of this class. You can find more information about this theorem online.

- **Polynomial Remainder Theorem.** The remainder of a polynomial $P(x)$ divided by a linear divisor $x - a$ is equal to $P(a)$.

- **Factor Theorem.** A polynomial $P(x)$ has a factor $x - a$ if and only if $P(a) = 0$.

- **Rational Root Theorem Theorem**
  - If $P(x)$ is a polynomial with leading coefficient a and constant term b, then any rational zero of $P(x)$ is of the form $\pm\dfrac{m}{n}$ where $m|b$ and $n|a$.
  - For example, the rational root theorem says that the only possible rational roots of $2x^5 + 3x^4 - 3x^3 - 3x^2 - 5x - 6 = 0$ are

$$\pm 1, \pm 2, \pm 3, \pm 6, \pm\frac{1}{2}, \pm\frac{3}{2}.$$

(In fact, the only rational roots are -2 and $\dfrac{3}{2}$.)

8.1 Example Questions

Problem 8.1 True or False. If the statement is false, explain how to correct the statement.

(a) If the degree of a polynomial $P(x)$ is d, then the number of terms of $P(x)$ is between 1 and d (inclusive).

(b) If the degrees of polynomials $p(y)$ and $q(y)$ are d and e, then the degree of $p(y) \cdot q(y)$ is $d + e$.

(c) If the degrees of polynomials $N(y)$ and $M(y)$ are d and e, then the degree of $N(M(y))$ is e^d.

Problem 8.2 For this problem, $f(x) = 3x + 2$, $g(x) = x - 7$, and $h(x) = x^2 - 4x + 4$. Compute the following values:

(a) $f(g(4))$

(b) $g(g(g(g(g(35)))))$

(c) $h(f(0))$

(d) $h(f(100))$

(e) $f(g(1234567)) - g(f(1234567))$

Problem 8.3 In the polynomial $(7+x)(1+x^2)(5+x^4)(2+x^8)(3+x^{16})(10+x^{32})$, what is the coefficient of x^{54}?

Problem 8.4 Compute the quotient and remainders of the following:

(a) $(x^5 + 4x^4 + 4x^3 + 11x^2 + 16x + 6) \div (x^3 + 2x + 3)$.

(b) $(x^5 + 4x^4 + 4x^3 + 11x^2 + 16x + 6) \div (x^3 + 2x^2 + 5)$.

Problem 8.5 Solve $\dfrac{x^4 + 4x^3 + 2x^2 - 4x + 5}{x^2 + 2x - 1} = 4$.

Problem 8.6 Prove the Polynomial Remainder Theorem

Problem 8.7 Solve $x^4 - 3x^3 - x^2 + 9x - 6 = 0$.

Problem 8.8 Expand $(x^2 - x + 1)^6$ to get $a_{12}x^{12} + a_{11}x^{11} + \cdots + a_1 x + a_0$. Find the value of $a_{12} + a_{10} + a_8 + a_6 + a_4 + a_2 + a_0$.

Problem 8.9 Let x be a real number such that $x^3 + 4x = 8$. Determine the value of $x^7 + 64x^2$.

Problem 8.10 Assume $(x-c)^2 \mid (4x^3 + 8x^2 - 11x + 3)$, find the value of c.

8.2 Quick Response Questions

Problem 8.11 Is a zero-degree polynomial the same as the zero polynomial?

Problem 8.12 Let $f(x) = 5x^3 - 2$ and $g(x) = x^2 + 2$. What is the degree of $f(g(x))$?

Problem 8.13 Let $f(x) = 5x^3 - 2$ and $g(x) = x^2 + 2$. What is the degree of $f(x) \times g(x)$?

Problem 8.14 Consider $x^3 - 6x^2 + 11x - 6 = 0$. Which of the following sets of roots are all POSSIBLE according to the Rational Root Theorem?

(A) $1, -4, 6$
(B) $0.5, 2, -6$
(C) $-1, 3, 6$
(D) $-2, -3, 12$

Problem 8.15 Find the product of all the real roots of $x^4 - 16 = 0$.

Problem 8.16 Find the remainder when $x^3 + 2x^2 + x + 1$ is divided by $x + 1$. Recall that since we are dividing by a linear equation our remainder is a constant (a number).

Problem 8.17 Find the remainder when $x^3 + 2x^2 + x + 1$ is divided by $x^2 + x + 1$.

(A) $x + 1$
(B) $-x$
(C) $x - 1$
(D) x

Problem 8.18 What is the constant term of $(x^3 + 2x + 1)(x + 2)^2(x^4 - 3x + 2)$.

Problem 8.19 What is the coefficient of x^8 in $(x^3 + 2x + 1)(x + 2)^2(x^4 - 3x + 2)$.

Problem 8.20 There is one real solution to $x^3 - 3x^2 + 4x - 2 = 0$. What is it? Round your answer to the nearest tenth if necessary.

8.3　Practice Questions

Problem 8.21 True or False. If the statement is false, explain how to correct the statement.

(a) If the degree of every term in the polynomial $g(x)$ is even, and every coefficient is positive, then $g(x) \geq 0$ for every possible real value of x.

(b) If the degree of a single term in the polynomial $g(x)$ is odd, then $g(x) = 0$ for some real value of x.

Problem 8.22 For this problem, $f(x) = 3x + 2$, $g(x) = x - 7$, and $h(x) = x^2 - 4x + 4$. Compute the following values:

(a) $g(f(4))$

(b) $h(f(1))$

Problem 8.23 What is the coefficient of x^4 in $(x^4 + 3)(x^2 + 3x + 4)(x^2 - 2x - 3)$?

Problem 8.24 Compute the quotient and remainder when $x^4 + 4x^2 + 3x + 2$ is divided by $x^2 - 4$.

Problem 8.25 Find all the real solutions of $\dfrac{x^5 - x^4 - 2x^3 + 3x^2 - 8x + 10}{x^2 + 2} = 1$.

Problem 8.26 Prove the Factor Theorem. Hint: Use the Polynomial Remainder Theorem.

Problem 8.27 Solve $x^4 - 4x^3 + x^2 + 4x - 2 = 0$.

Problem 8.28 Assume $(3x - 1)^7 = a_7 x^7 + a_6 x^6 + \cdots + a_1 x + a_0$, find the value of $a_0 + a_2 + a_4 + a_6$.

Problem 8.29 Let x be a real number such that $x^3 + 4x = 8$. Determine the value of $x^7 + 64x^2$.

Problem 8.30 Find all possible rational c such that $(x - c) \mid 2x^5 - 3x^4 - 2x^3 - 2x^2 + 3x + 2$.

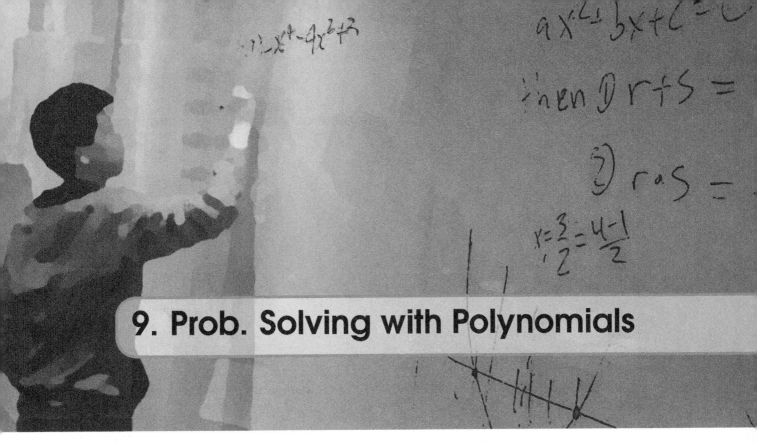

9. Prob. Solving with Polynomials

Useful Theorems

- **Polynomial Remainder Theorem**. The remainder of a polynomial $P(x)$ divided by a linear divisor $x - a$ is equal to $P(a)$.
- **Factor Theorem**. A polynomial $P(x)$ has a factor $x - a$ if and only if $P(a) = 0$.
- **Vieta's Theorem (quadratic version)**. Let x_1, x_2 be the roots of a quadratic equation $ax^2 + bx + c = 0$, where $a \neq 0$, then

$$x_1 + x_2 = -\frac{b}{a}, \qquad \text{and} \qquad x_1 x_2 = \frac{c}{a}.$$

- **Vieta's Theorem (general version)**. Let $P(x) = a_n x^n + a_{n-1} x^{n-1} + \cdots + a_0$ be a polynomial of degree n, and $x_1, x_2, \ldots, x_n$ be the zeros of $P(x)$. Then

$$\begin{cases} x_1 + x_2 + \cdots + x_n &= -\dfrac{a_{n-1}}{a_n}, \\ x_1 x_2 + x_1 x_3 + \cdots + x_{n-1} x_n &= \dfrac{a_{n-2}}{a_n}, \\ &\cdots \\ x_1 x_2 \cdots x_n &= (-1)^n \dfrac{a_0}{a_n}. \end{cases}$$

9.1   Example Questions

Problem 9.1 Let $m \geq -1$ be a real number, and the equation $x^2 + 2(m-2)x + m^2 - 3m + 3 = 0$ has two distinct real roots x_1 and x_2. If $x_1^2 + x_2^2 = 6$, what is m?

Problem 9.2 Let a, b, c, and d be the roots of $x^4 - 2x - 1990 = 0$. Find the value of $1/a + 1/b + 1/c + 1/d$.

Problem 9.3 An $l \times w \times h$ rectangular box has surface area 38 and volume 12. If $l + w + h = 8$, find the dimensions of the box.

Problem 9.4 Find the sum of the 17th powers of the 17 roots of $x^{17} - 3x + 1 = 0$.

Problem 9.5 Distinct real numbers a and b satisfies $(a+1)^2 = 3 - 3(a+1), 3(b+1) = 3 - (b+1)^2$. Find the value of $b\sqrt{\dfrac{b}{a}} + a\sqrt{\dfrac{a}{b}}$.

Problem 9.6 Find ordered pairs (x, y) of real numbers such that $x^2 - xy + y^2 = 13$ and $x - xy + y = -5$.

Problem 9.7 If $x + y + z = 0$ and $x^3 + y^3 + z^3 = 288$, find the value of xyz.

Problem 9.8 The polynomial $p(x) = x^3 + 2x^2 - 5x + 1$ has three different roots a, b, and c. Find $a^3 + b^3 + c^3$.

Problem 9.9 Let $x = \dfrac{2}{2+\sqrt{3}-\sqrt{5}}, y = \dfrac{2}{2+\sqrt{3}+\sqrt{5}}$, evaluate:

$$\frac{x^4 y^4}{x^4 + y^4 + 6x^2 y^2 + 4x^3 y + 4xy^3}.$$

Problem 9.10 Let x and y be nonzero real numbers satisfying $|x| + y = 3$ and $|x|y + x^3 = 0$, Find the value of $x + y$.

9.2   Quick Response Questions

Problem 9.11 Factor $x^3 + x^2y^2 + x^2y + xy^3$.

(A) $x(x+y)(x^2+y^2)$
(B) $x(x^2+y)(x+y^2)$
(C) $(x+y)(x+y^2)$
(D) $x(x+y)(x+y^2)$

Problem 9.12 Using the rational root theorem, find all the rational roots of $x^4 - x^3 - 5x^2 + 3x + 2 = 0$.

(A) $1, 2$
(B) $1, -2$
(C) $1, -1, -2$
(D) $-1, -2$

Problem 9.13 There are two irrational roots of $x^4 - x^3 - 5x^2 + 3x + 2 = 0$, with the form $A \pm \sqrt{B}$ for integers A, B. What is $A + B$?

Problem 9.14 Let

$$f(x) = x^4 + x^3 - 11x^2 - 5x + 30 = (x-2)(x+3)(x^2-5).$$

Verify Vieta's formula for the sum of the roots. What is the sum of the roots?

Problem 9.15 Let

$$f(x) = x^4 + x^3 - 11x^2 - 5x + 30 = (x-2)(x+3)(x^2-5).$$

Verify Vieta's formula for the product of the roots. What is the product of the roots?

Problem 9.16 Let

$$f(x) = x^4 + x^3 - 11x^2 - 5x + 30 = (x-2)(x+3)(x^2-5),$$

and denote the four roots a, b, c, d. Verify Viete's formula for $abc + abd + acd + bcd$. What is $abc + abd + acd + bcd$

Problem 9.17 What is the sum of the roots of $x^5 + 2x^4 + 3x^3 + 4x^2 + 5x + 6 = 0$?

Problem 9.18 Find the sum of the 5th powers of the 5 roots of $x^5 + 2x - 1 = 0$.

Problem 9.19 How many real roots does $x^6 - 1 = 0$ have?

Problem 9.20 Suppose $P(x) = ax^3 + bx^2 + cx + d$. If we are given that $P(0) = 2$, $P(1) = 3$, and $P(-1) = 1$, what is b?

9.3 Practice Questions

Problem 9.21 The two roots of equation $x^2 + px + 1 = 0$, where $p > 0$, have difference 1. Find the value of p.

Problem 9.22 Let a, b, c be the roots of $2x^3 + 20x^2 - 75x + 50 = 0$. Find the value of $\dfrac{1}{ab} + \dfrac{1}{ac} + \dfrac{1}{bc}$.

Problem 9.23 The volume of a box is 100 and the surface area is 160. Given that one of the sides is 2, what is the sum of all three dimensions? Try to use Viete's in your solution.

Problem 9.24 Find the sum of the 5th powers of the 5 roots of $x^5 + 2x - 1$.

Problem 9.25 If a, b, c, d are four different numbers for which

$$\begin{cases} a^4 + a^2 + ka + 64 = 0 \\ b^4 + b^2 + kb + 64 = 0 \\ c^4 + c^2 + kc + 64 = 0 \\ d^4 + d^2 + kd + 64 = 0. \end{cases}$$

What is the value of $a^2 + b^2 + c^2 + d^2$? Hint: Consider the polynomial $x^4 + x^2 + kx + 64$.

Problem 9.26 Find all pairs (x, y) of real numbers so that $x + xy + y = 4$ and $x^2 + xy + y^2 = 8$.

Problem 9.27 Prove the factoring formula $x^3 + y^3 + z^3 - 3xyz = (x+y+z)(x^2 + y^2 + z^2 - xy - yz - zx)$ using the following steps.

(a) Rewrite $x^3 + y^3 + z^3 - 3xyz$ as a polynomial $P(z)$ (that is, think of x, y as constants and write as a polynomial in z).

(b) Show that $-(x+y)$ is a root of $P(z) = 0$.

(c) By part (b) and the factor theorem, we know $(z+x+y)$ is a factor of $P(z)$. Use this to complete the factorization.

Problem 9.28 Suppose that the roots of $3x^3 + 3x^2 + 4x - 11 = 0$ are a, b and c, and the roots of $x^3 + rx^2 + sx + t = 0$ are $a+b, b+c$, and $c+a$. Find t.

Problem 9.29 Let $x, y = 3 \pm \sqrt{5}$. Calculate $\dfrac{x^2 + xy + y^2}{x^3y^2 + x^2y^3 + x + y}$.

Problem 9.30 Solve the system of equations $|x| + y = 2$ and $x + y^2 = 4$.

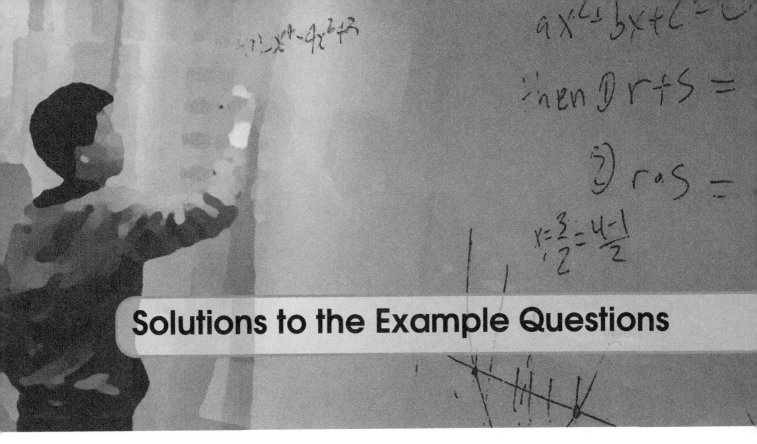

Solutions to the Example Questions

In the sections below you will find solutions to all of the Example Questions contained in this book.

Quick Response and Practice questions are meant to be used for homework, so their answers and solutions are not included. Teachers or math coaches may contact Areteem at info@areteem.org for answer keys and options for purchasing a Teachers' Edition of the course.

1 Solutions to Chapter 1 Examples

Problem 1.1 Factor the following.

(a) $10x^{10}y^8 + 5x^5y^9$.

Answer

$5x^5y^8(2x^5 + y)$

Solution

The constants share a common factor of 5, the x terms share a common factor of x^5, and the y terms share a common factor of y^8. Dividing the factor $5x^5y^8$ from both addends yields $2x^5 + y$.

(b) $(2x + 3y)^3 - 8x^3 - 27y^3$.

Answer

$18xy(2x + 3y)$

Solution

Expand $(2x + 3y)^3 = 8x^3 + 36x^2y + 54xy^2 + 27y^3$. Cancel like terms and factor $36x^2y + 54xy^2$. The constants share a common factor of 18, the x terms share a common factor of x and the y terms share a common factor of y. Dividing the factor $18xy$ from both addends yields $2x + 3y$.

Problem 1.2 Prove the following identities:

(a) Prove the formula for factoring $a^3 + b^3 + c^3 - 3abc$.

Solution

$$\begin{aligned} a^3 + b^3 + c^3 - 3abc &= (a+b)^3 - 3ab(a+b) + c^3 - 3abc \\ &= (a+b)^3 + c^3 - 3ab(a+b+c) \\ &= (a+b+c)[(a+b)^2 - c(a+b) + c^2] - 3ab(a+b+c) \\ &= (a+b+c)(a^2 + b^2 + c^2 - ab - bc - ca) \end{aligned}$$

A second method to prove this formula would be to expand the factored form and see if the result is identical to the other side.

(b) Prove the formula for factoring $y^6 - y^3$.

Solution

$$\begin{aligned} y^6 - y^3 &= y^3(y^3 - 1) \\ &= y^3(y - 1)(y^2 + y + 1) \end{aligned}$$

We could also do $y^6 - y^3 = (y^2)^3 - y^3$ and apply the difference of cubes formula. However, finding common factors first is usually simpler.

Problem 1.3 Factor $x^{15} + x^{14} + x^{13} + \cdots + x^2 + x + 1$.

Answer

$(x^8 + 1)(x^4 + 1)(x^2 + 1)(x + 1)$

Solution

Note that $(x - 1)(x^{15} + x^{14} + x^{13} + \cdots + x^2 + x + 1) = x^{16} - 1 = (x^8 + 1)(x^8 - 1) = (x^8 + 1)(x^4 + 1)(x^4 - 1) = (x^8 + 1)(x^4 + 1)(x^2 + 1)(x^2 - 1) = (x^8 + 1)(x^4 + 1)(x^2 + 1)(x + 1)(x - 1)$. Dividing both sides by $(x - 1)$ yields the desired result.

Problem 1.4 Factor the following.

(a) Factor $22y^2 - 35y + 3$.

Answer

$(2y - 3)(11y - 1)$

Solution

Use some trial and error to get $22 = 2 \times 11$ and $3 = (-3)(-1)$, and $-35 = 2(-1) + 11(-3)$, thus

$$22y^2 - 35y + 3 = (2y - 3)(11y - 1).$$

This factors the equation as needed.

(b) Factor $2x^2 - 7x + 3$.

Answer

$(2x-1)(x-3)$.

Solution

Note that the leading term's coefficient 2 can be factored into 2 and 1 and the constant term 3 can be factored into -3 and -1. Apply cross multiplication to determine that $2(-3)+1(-1)=-7$ is the middle term's coefficient. Therefore, $2x^2-7x+3=(2x-1)(x-3)$.

(c) Factor $3k^2-5k-2$.

Answer

$(3k+1)(k-2)$.

Solution

Note that the leading term's coefficient 3 can be factored into 3 and 1 and the constant term -2 can be factored into -2 and 1. Apply cross multiplication to determine that $3(-2)+1(1)=-5$ is the middle term's coefficient. Therefore, $3k^2-5k-2=(3k+1)(k-2)$.

Problem 1.5 Factor $a^4+a^3+a^2b+ab^2+b^3-b^4$.

Answer

$(a+b)(a^2+b^2)(a-b+1)$

Solution

$$\begin{aligned}
&a^4+a^3+a^2b+ab^2+b^3-b^4\\
=\ &(a^4-b^4)+(a^3+b^3)+(a^2b+ab^2)\\
=\ &(a+b)(a-b)(a^2+b^2)+(a+b)(a^2-ab+b^2)+(a+b)ab\\
=\ &(a+b)[(a^2+b^2)(a-b)+(a^2-ab+b^2)+ab]\\
=\ &(a+b)[(a^2+b^2)(a-b)+(a^2+b^2)]\\
=\ &(a+b)(a^2+b^2)(a-b+1).
\end{aligned}$$

Problem 1.6 Factor x^3-9x+8 by using the given technique:

(a) $8 = -1 + 9$.

Solution

$x^3 - 9x + 8 = x^3 - 9x - 1 + 9 = (x^3 - 1) - 9(x - 1) = (x - 1)(x^2 + x + 1) - 9(x - 1) = (x - 1)(x^2 + x + 1 - 9) = (x - 1)(x^2 + x - 8)$.

(b) $-9x = -x - 8x$.

Solution

$x^3 - 9x + 8 = x^3 - x - 8x + 8 = (x^3 - x) - (8x - 8) = x(x + 1)(x - 1) - 8(x - 1) = (x^2 + x - 8)(x - 1)$.

(c) $x^3 = 9x^3 - 8x^3$.

Solution

$x^3 - 9x + 8 = 9x^3 - 8x^3 - 9x + 8 = (9x^3 - 9x) - (8x^3 - 8) = 9x(x + 1)(x - 1) - 8(x - 1)(x^2 + x + 1) = (x - 1)(9x^2 + 9x - 8x^2 - 8x - 8) = (x - 1)(x^2 + x - 8)$.

(d) Add two terms: $-x^2 + x^2$.

Solution

$x^3 - 9x + 8 = x^3 - x^2 + x^2 - 9x + 8 = (x^3 - x^2) + (x^2 - 9x + 8) = x^2(x - 1) + (x - 8)(x - 1) = (x^2 + x - 8)(x - 1)$.

Problem 1.7 Factor the following.

(a) $x^9 + x^6 + x^3 - 3$.

Answer

$(x - 1)(x^2 + x + 1)(x^6 + 2x^3 + 3)$.

Solution

If we re-express $-3 = -1 - 1 - 1$, we have

$$
\begin{aligned}
x^9 + x^6 + x^3 - 3 &= (x^9 - 1) + (x^6 - 1) + (x^3 - 1) \\
&= (x^3 - 1)(x^6 + x^3 + 1) + (x^3 - 1)(x^3 + 1) + (x^3 - 1) \\
&= (x^3 - 1)[(x^6 + x^3 + 1) + (x^3 + 1) + 1] \\
&= (x - 1)(x^2 + x + 1)(x^6 + 2x^3 + 3).
\end{aligned}
$$

(b) (Sophie Germain's technique) $x^4 + 4y^4$.

Answer

$(x^2 + 2xy + 2y^2)(x^2 - 2x^2 y^2 + 2y^2)$.

Solution

$x^4 + 4y^4 = x^4 + 4x^2 y^2 + 4y^4 - 4x^2 y^2 = (x^2 + 2y^2)^2 - (2xy)^2 = (x^2 + 2xy + 2y^2)(x^2 - 2x^2 y^2 + 2y^2)$.

Problem 1.8 Factor the expression $(x+1)^4 + (x^2 - 1)^2 + (x - 1)^4$.

Answer

$(3x^2 + 1)(x^2 + 3)$

Solution

If we add and subtract $(x-1)^2(x+1)^2$, we have

$$
\begin{aligned}
&\quad (x+1)^4 + (x^2 - 1)^2 + (x - 1)^4 \\
&= (x+1)^4 + 2(x+1)^2(x-1)^2 + (x-1)^4 - (x-1)^2(x+1)^2 \\
&= ((x+1)^2 + (x-1)^2)^2 - (x^2 - 1)^2 \\
&= (2x^2 + 2)^2 - (x^2 - 1)^2 \\
&= (2x^2 + 2 + x^2 - 1)(2x^2 + 2 - (x^2 - 1)) \\
&= (3x^2 + 1)(x^2 + 3).
\end{aligned}
$$

Problem 1.9 Factor $a^2 + b^2 + c^2 - 2bc + 2ca - 2ab$.

Answer

$(a-b+c)^2$

Solution

Recall that $(a+b+c)^2 = a^2+b^2+c^2+2ab+2bc+2ca$. Therefore, $(a+(-b)+c)^2 = a^2+(-b)^2+c^2+2a(-b)+2(-b)c+2ca = a^2+b^2+c^2-2ab-2bc+2ca$ and hence our expression factors as $(a-b+c)^2$.

Problem 1.10 Factor $-2x^{5n-1}y^n+4x^{3n-1}y^{n+2}-2x^{n-1}y^{n+4}$.

Answer

$-2x^{n-1}y^n(x^n-y)^2(x^n+y)^2$

Solution

The constants share a common factor of -2, the x terms share a common factor of x^{n-1}, and the y terms share a common factor of y^n. Dividing the factor $-2x^{n-1}y^n$ from both addends yields $x^{4n}-2x^{2n}y^2+y^4$. Note that, $x^{4n}-2x^{2n}y^2+y^4 = (x^{2n})^2-2x^{2n}y^2+(y^2)^2 = (x^{2n}-y^2)^2$. Furthermore, note that, $x^{2n}-y^2 = (x^n)^2-(y)^2 = (x^n-y)(x^n+y)$. Combining the identities yields the desired solution.

2 Solutions to Chapter 2 Examples

Problem 2.1 Determine the number of roots without solving the equation:

(a) $3x^2 + 4x - 5 = 0$

Answer

Two real roots.

Solution

$\Delta = 4^2 - 4(3)(-5) = 16 + 60 = 76 > 0$. Thus there are two real roots.

(b) $4x^2 + 20x + 25 = 0$

Answer

One real root.

Solution

$\Delta = 20^2 - 4(4)(25) = 400 - 400 = 0$. Hence there is one real (double) root.

(c) $2x^2 + 2x + 3 = 0$

Answer

No real roots.

Solution

$\Delta = 2^2 - 4(2)(3) = 4 - 24 = -20 < 0$. Therefore there are no real roots.

Problem 2.2 For each of the following, state whether the function has a maximum or a minimum and find the maximum or minimum value.

(a) $f(x) = 4x^2 - 12x - 16$.

Answer

Minimum: -25.

Solution

The vertex is $-(-12)/(2 \cdot 4) = 3/2$ and the x^2 coefficient is positive, hence $f(3/2) = -25$ is a minimum.

(b) $f(x) = 61 + 72x - 36x^2$.

Answer

Maximum: 97.

Solution

The vertex is $-72/(2 \cdot (-36)) = 1$ and the x^2 coefficient is negative, hence $f(1) = 97$ is a maximum.

(c) $f(x) = x^4 - 4x^2 + 3$.

Answer

Minimum: -1.

Solution

First, note that $g(u) = u^2 - 4u + 3$ has the minimum value $g(2) = -1$ at the vertex $-(-4)/(2 \cdot 1) = 2$. Since $f(x) = g(x^2)$, $f(x)$ assumes its minimum value of -1 at $x = \pm\sqrt{2}$.

Problem 2.3 For what x are the following functions positive?

(a) $f(x) = 4x^2 - 12x - 16$.

Answer

$x > 4$ or $x < -1$.

Solution

First, show by factoring or the quadratic formula that the roots of $fx) = 0$ are $x = -1, 4$. Now $f(0) = -16 < 0$ implies that $f(x) \leq 0$ when $-1 \leq x \leq 4$, so $f(x)$ is positive on the rest: $x > 4$ or $x < -1$.

(b) $f(x) = -x^2 + 4x + 2$.

Answer

$2 - \sqrt{6} < x < 2 + \sqrt{6}$.

Solution

Use the quadratic formula to show that $fx) = 0$ when $x = 2 \pm \sqrt{6}$. Now $f(2) = 6 > 0$ implies that $f(x) > 0$ when $2 - \sqrt{6} < x < 2 + \sqrt{6}$.

(c) $f(x) = 2x^2 + 2x + 3$

Answer

All real numbers.

Solution

Here the vertex is $-2/(2 \cdot 2) = -1/2$ and the x^2 coefficient is positive, so $f(-1/2) = 5/2 > 0$ is the minimum. Hence $f(x)$ is always positive.

(d) $f(x) = x^4 - 4x^2 + 3$.

Answer

$|x| > \sqrt{3}$ or $|x| < 1$.

Solution

Since $u^2 - 4u + 3 = (u - 1)(u - 3)$ has roots at $u = 1, 3$ and $g(2) = -1 < 0$, $g(u)$ is positive for $u < 1$ or $u > 3$. Substituting $u = x^2$ gives $x^2 < 1$ or $x^2 > 3$. Hence $f(x)$ is positive when $|x| < 1$ or $|x| > \sqrt{3}$.

Problem 2.4 Given that the equation $x^2 - 2x - m = 0$ has no real roots, how many real roots does the equation $x^2 + 2mx + 1 + 2(m^2 - 1)(x^2 + 1) = 0$ have?

Answer

None.

Solution

Derive $m < -1$ from $\Delta_1 < 0$, and $\Delta_2 = -4(2m-1)(m+1)(2m+1)(2m-1) < 0$. Or show the left hand side is always positive.

Problem 2.5 For what values of m does the equation $4x^2 + 8x + m = 0$ have two distinct real roots?

Answer

$m < 4$

Solution

$\Delta = 8^2 - 4 \cdot 4m = 64 - 16m > 0$, so $m < 4$.

Problem 2.6 Use Vieta's formula to solve the following.

(a) A quadratic equation has two roots $\dfrac{2}{3}$ and $-\dfrac{1}{2}$, what is this equation? (multiple answers are possible)

Answer

One example: $6x^2 - x - 2$

Solution

We have
$$\frac{2}{3} - \frac{1}{2} = \frac{1}{6} = -\frac{b}{a}, \frac{2}{3} \cdot -\frac{1}{2} = -\frac{1}{3} = \frac{c}{a}.$$
Thus the equation is $x^2 - \dfrac{1}{6}x - \dfrac{1}{3} = 0$, which can be re-written as $6x^2 - x - 2 = 0$. (Or any equation with coefficients proportional to this one.

(b) Two real numbers have sum -10 and product -5, find these two numbers.

Answer

$-5 + \sqrt{30}; -5 - \sqrt{30}$

Solution

These two numbers are the roots of the equation

$$x^2 + 10x - 5 = 0.$$

Solving to get $x = \dfrac{-10 \pm \sqrt{10^2 + 4(5)}}{2} = -5 \pm \sqrt{30}$. So the two numbers are $-5 + \sqrt{30}$ and $-5 - \sqrt{30}$.

Problem 2.7 Answer the following.

(a) (2008 AMC 10B) A quadratic equation $ax^2 - 2ax + b = 0$ has two real solutions. What is the average of the solutions?

Answer

1.

Solution

By Vieta's Theorem, $x_1 + x_2 = \dfrac{2a}{a} = 2$, So the average $\dfrac{x_1 + x_2}{2} = 1$.

(b) (2005 AMC 10B/12B) The quadratic equation $x^2 + mx + n = 0$ has roots that are twice those of $x^2 + px + m = 0$, and none of m, n, and p is zero. What is the value of n/p?

Answer

8.

Solution

Assume the two roots of the latter equation are x_1 and x_2, then the roots of the former equation are $2x_1$ and $2x_2$. By Viète' Theorem, $x_1 + x_2 = -p$, $x_1 x_2 = m$, and also $2x_1 + 2x_2 = -m$ and $(2x_1)(2x_2) = n$. So we get $m = 2p, 4m = n$. So $n/p = 8$.

Problem 2.8 Let x_1, x_2 be the two roots for equation $x^2 + x - 3 = 0$, find the value of $x_1^3 - 4x_2^2 + 19$.

Answer

0.

Solution

x_1, x_2 are roots of the equation $x^2 + x - 3$, so $x_1^2 + x_1 - 3 = 0$ and similarly $x_2^2 + x_2 - 3 = 0$. Hence, $x_1^2 = 3 - x_1, x_2^2 = 3 - x_2$, and we have $x_1^3 - 4x_2^2 + 19 = x_1(3 - x_1) - 4(3 - x_2) + 19$. Simplifying and substituting again we have $3x_1 - x_1^2 + 4x_2 + 7 = 3x_1 - (3 - x_1) + 4x_2 + 7 = 4(x_1 + x_2) + 4$. By Vieta's Theorem we have $x_1 + x_2 = -1$, so $4(x_1 + x_2) + 4 = 4(-1) + 4 = 0$.

Problem 2.9 Given the equation in x, $x^2 + 2mx + m + 2 = 0$:

(a) For what values of m does the equation have two (not necessarily distinct) positive roots?

Answer

$-2 < m \le -1$

Solution

First note for the equation to have two real roots we have $\Delta = (2m)^2 - 4 \cdot (m+2) = 4m^2 - 4m - 8 = 4(m-2)(m+1) \ge 0$, so $m \ge 2$ or $m \le -1$.

If the roots are x_1, x_2, $x_1, x_2 > 0$ means that $x_1 + x_2 > 0$ and $x_1 \cdot x_2 > 0$. Thus by Vieta's Theorem we have (sum of roots) $-2m > 0$ and (product of roots) $m + 2 > 0$. Hence $m < 0$ and $m > -2$.

We therefore see that $m \ge 2$ is impossible, so $m \le -1$, $m < 0$, and $m > -2$. hence $-2 < m \le -1$.

(b) For what values of m does the equation have one positive root and one negative root?

Answer

$m < -2$.

Solution

Similar to part (a), $\Delta \ge 0$ implies $m \ge 2$ or $m \le -1$.

If the roots are x_1, x_2 with $x_1 < 0 < x_2$, we have $x_1 \cdot x_2 < 0$ (note we cannot say whether

$x_1 + x_2$ is positive or negative). By Vieta's theorem, (product of roots) $m + 2 < 0$, so $m < -2$. Hence we must have $m \leq -1$ and $m < -2$, so $m < -2$.

Problem 2.10 (2005 AMC 10A/12A) There are two values of a for which the equation $4x^2 + ax + 8x + 9 = 0$ has only one solution for x. What is the sum of those values of a?

Answer

-16

Solution

For the quadratic equation to have only one solution, the discriminant has to be 0. Thus $\Delta = (a + 8)^2 - 4 \cdot 4 \cdot 9 = 0$. Therefore $(a + 8)^2 = 144$, and then $a + 8 = \pm 12$, which means $a = 4$ or $a = -20$. So the answer is $4 + (-20) = -16$.

3 Solutions to Chapter 3 Examples

Problem 3.1 Answer the following.

(a) How many real roots does the equation

$$(7x - 13)(x^2 + 7x - 13)(x^2 - 7x + 13)$$

have?

Answer

3

Solution

A root must satisfy $7x - 13 = 0$, $x^2 + 7x - 13 = 0$, or $x^2 - 7x + 13 = 0$. The first equation has one real solution, the second equation has two real solutions since the discriminant $7^2 - 4(1)(-13) = 101 > 0$, and the third equation has no real solutions since the discriminant $(-7)^2 - 4(1)(13) = -3 < 0$.

(b) Given $p > 0$ and $q < 0$, how many positive roots does the equation $x^2 + px + q = 0$ have?

Answer

1.

Solution

$\Delta = p^2 - 4q > 0$ means there are two real roots, and $x_1 x_2 = q < 0$ means one is positive and one is negative.

(c) Without solving the equation, find the number of real roots for x: $(n^2 + 1)x^2 - 2nx + (n^2 + 4) = 0$.

Answer

There are no real roots.

Solution

$\Delta = (2n)^2 - 4(n^2 + 1)(n^2 + 4) = 4n^2 - 4n^4 - 20n^2 - 16 = -4(n^2 + 2)^2 < 0.$

Problem 3.2 Find k in each of the following scenarios:

(a) In the equation $x^2 - 402x + k = 0$, one of the roots plus three equals 80 times the other root.

Answer

$k = 1985$

Solution

The two roots are 397 and 5.

(b) Let x_1 and x_2 be the two roots of the equation $4x^2 - 8x + k = 0$. Suppose further that $\dfrac{1}{x_1} + \dfrac{1}{x_2} = \dfrac{8}{3}$.

Answer

$k = 3$

Solution

By Vieta's formulas, $x_1 + x_2 = 2, x_1 x_2 = \dfrac{k}{4}$. Thus

$$\frac{1}{x_1} + \frac{1}{x_2} = \frac{x_1 + x_2}{x_1 x_2} = \frac{8}{k} = \frac{8}{3}.$$

Therefore $k = 3$.

Problem 3.3 Do the following.

(a) The sum of squares of the roots of equation $x^2 + 2kx = 3$ is 10. Find the possible values of k.

Answer

± 1

Solution

Let x_1 and x_2 be the roots of the equation. Applying Vieta's Theorem to the quadratic equation $x^2 + 2kx - 3 = 0$ yields $x_1 + x_2 = -2k$ and $x_1 x_2 = -3$. Note that, $4k^2 =$

$(-2k)^2 = (x_1 + x_2)^2 = x_1^2 + 2x_1x_2 + x_2^2 = 10 + 2(-3) = 4$. This implies that $k = \pm 1$.

(b) For equation $2x^2 + mx - 2m + 1 = 0$, the sum of squares of the two real roots is $\dfrac{29}{4}$. Find the value of m.

Answer

$m = 3$.

Solution

Use Vieta's Theorem, $x_1^2 + x_2^2 = (x_1 + x_2)^2 - 2x_1x_2 = \left(\dfrac{m}{2}\right)^2 - (-2m + 1) = \dfrac{29}{4}$. Solve for m to get $m = 3$ and $m = -11$. The other solution $m = -11$ doesn't provide real roots for the equation, so throw away.

Problem 3.4 The quadratic equation $x^2 + 2kx + 2k^2 - 1 = 0$ has at least one negative root. Find the possible range of values for k.

Answer

$-\sqrt{2}/2 < k \leq 1$.

Solution

The equation has real roots if $-1 \leq k \leq 1$, and has two non-negative roots if $-1 \leq k < -\sqrt{2}/2$.

Problem 3.5 The two real roots of $x^2 + (m - 2)x + 5 - m = 0$ are both greater than 2. Find the possible range of values for real number m.

Answer

$-5 < m \leq -4$.

Solution

Note that $x^2 + (m - 2)x + 5 - m = (x - 2)^2 + (m + 2)(x - 2) + (m + 5)$. Since there exists two real roots, the discriminant must be nonnegative. Equivalently, $(m + 2)^2 - 4(m + 5) = m^2 + 4m + 4 - 4m - 20 = m^2 - 16 \geq 0$. Therefore $m \leq -4$. Applying Vieta's

Theorem yields $(x_1 - 2)(x_2 - 2) = m + 5$. Given that the real roots must also be greater than 2, we have $(x_1 - 2)(x_2 - 2) > 0$. Therefore, $m + 5 > 0$ or $m > -5$.

Problem 3.6 Find all ordered pairs (a, b) such that $a^2 + b^2$ is prime, and the equation $x^2 + ax + 1 = b$ has two positive integer roots.

Answer

No such pairs exist.

Solution

Let x_1 and x_2 be the two positive integer roots, then $x_1 + x_2 = -a$ and $x_1 x_2 = 1 - b$, thus $a^2 + b^2 = (x_1 + x_2)^2 + (1 - x_1 x_2)^2 = (x_1^2 + 1)(x_2^2 + 1)$ must be a composite number.

Problem 3.7 The equation $x^2 + (a - 6)x + a = 0$ has two integer roots. Find the value of a.

Answer

0 or 16.

Solution

$x_1 + x_2 = 6 - a$ and $x_1 x_2 = a$, so $x_1 + x_2 + x_1 x_2 + 1 = 7$. Factoring, get the two roots are 0 and 6, or -8 and -2.

Problem 3.8 More Vieta Practice.

(a) If x_1 and x_2 are the two real roots of $x^2 + x + q = 0$, and $|x_1 - x_2| = q$, find the value of q.

Answer

$q = \sqrt{5} - 2$.

Solution

To solve, note that $q^2 = |x_1 - x_2|^2 = (x_1 + x_2)^2 - 4x_1 x_2 = 1 - 4q$. Solve to find the positive value of q.

(b) For the equation $x^2 + mx + n = 0$, the difference between the two roots is p and the product of the two roots is q. What is $m^2 + n^2$ in terms of p and q?

Answer

$p^2 + 4q + q^2$

Solution

Let x_1 and x_2 be the roots of the equation. Applying Vieta's Theorem to the quadratic equation $x^2 + mx + n = 0$ yields $x_1 + x_2 = -m$ and $x_1 x_2 = n = q$. Note that $m^2 = (-m)^2 = (x_1 + x_2)^2 = x_1^2 + 2x_1 x_2 + x_2^2 = x_1^2 - 2x_1 x_2 + x_2^2 + 4x_1 x_2 = (x_1 - x_2)^2 + 4x_1 x_2 = p^2 + 4q$. Therefore, $m^2 + n^2 = p^2 + 4q + q^2$.

Problem 3.9 Find all real solutions to the system of equations: $x + y = 2$ and $xy - z^2 = 1$. Justify your answer.

Answer

$x = 1, y = 1, z = 0$.

Solution

Construct equation $t^2 - 2t + (z^2 + 1) = 0$, $\Delta \geq 0$ requires $z = 0$.

Problem 3.10 If x_1 and x_2 are integer roots of the equation $x^2 + mx + 2 - n = 0$, and $(x_1^2 + 1)(x_2^2 + 1) = 10$, how many possible pairs (m, n) are there?

Answer

6.

Solution

Since x_1 and x_2 are integers, the possible values for $x_1^2 + 1$ and $x_2^2 + 1$ are $1, 10$ and $2, 5$. In the first case, $x_1 = 0, x_2 = \pm 3$; and in the second case, $x_1 = \pm 1$ and $x_2 = \pm 2$. The order of x_1 and x_2 does not matter, since we only need to find the possible pairs of (m, n). By Viete's formulas, $-m = x_1 + x_2, 2 - n = x_1 x_2$, so there are 6 different pairs for (m, n).

4   Solutions to Chapter 4 Examples

Problem 4.1 Factor the following.

(a) $(x^2+x+1)(x^2+x+2)-12$. Hint: Try letting $y=x^2+x+1$.

Answer

$(x-1)(x+2)(x^2+x+5)$.

Solution

Let $y=x^2+x+1$, the expression becomes $y(y+1)-12=y^2+y-12=(y+4)(y-3)$, then change back to x.

(b) $(x^2+3x+2)(x^2+7x+12)-120$. Hint: Factor and regroup so you can make the substitution x^2+5x+5.

Answer

$(x^2+5x+16)(x+6)(x-1)$.

Solution

$(x^2+3x+2)(x^2+7x+12)-120=(x+1)(x+2)(x+3)(x+4)-120=(x^2+5x+4)(x^2+5x+6)-120$. Let $y=x^2+5x+5$, then $(y-1)(y+1)-120=y^2-121=(y+11)(y-11)$. We can then change back to x.

Problem 4.2 Factor the following using a change of variables.

(a) $x^2+x-14-\dfrac{1}{x}+\dfrac{1}{x^2}$. Hint: Note $\left(x-\dfrac{1}{x}\right)^2=x^2-2+\dfrac{1}{x^2}$.

Answer

$\left(x-\dfrac{1}{x}-3\right)\left(x-\dfrac{1}{x}+4\right)$.

Solution

We an rewrite the equation as

$$x^2-2+\frac{1}{x^2}+x-\frac{1}{x}-12,$$

so after the substitution $y = x - \dfrac{1}{x}$ we have $y^2 + y - 12 = (y+4)(y-3)$. Rewriting in terms of x gives the final answer.

(b) $6x^4 + 7x^3 - 36x^2 - 7x + 6$.

Answer

$(2x+1)(x-2)(3x-1)(x+3)$

Solution

$6x^4 + 7x^3 - 36x^2 - 7x + 6 = x^2 \left(6x^2 + 7x - 36 - \dfrac{7}{x} + \dfrac{6}{x^2} \right)$. Let $y = x - \dfrac{1}{x}$, $y^2 = x^2 - 2 + \dfrac{1}{x^2}$, thus
$6x^4 + 7x^3 - 36x^2 - 7x + 6 = x^2(6(y^2 + 2) + 7y - 36) = x^2(2y - 3)(3y + 8) = (2xy - 3x)(3xy + 8x) = (2x^2 - 3x - 2)(3x^2 + 8x - 3) = (2x+1)(x-2)(3x-1)(x+3)$

Problem 4.3 Factor $(x+3)(x^2 - 1)(x+5) - 20$

Answer

$(x^2 + 4x - 7)(x^2 + 4x + 5)$

Solution

We have $(x+3)(x^2 - 1)(x+5) - 20 = (x+3)(x+1)(x-1)(x+5) - 20 = (x^2 + 4x + 3)(x^2 + 4x - 5) - 20$, then let $y = x^2 + 4x$. We thus have $(y+3)(y-5) - 20 = y^2 - 2y - 35 = (y-7)(y+5)$. Substituting back x gives the final answer $(x^2 + 4x - 7)(x^2 + 4x + 5)$.

Problem 4.4 Factor $(x^2 + xy + y^2)^2 - 4xy(x^2 + y^2)$. Hint: Let $u = x+y, v = xy$.

Answer

$(x^2 - xy + y^2)^2$

Solution

Making the subsitution, note $x^2 + y^2 = (x+y)^2 - 2xy = u^2 - 2v$. Thus the equation becomes

$$(u^2 - v)^2 - 4v(u^2 - 2v) = u^4 - 2u^2v + v^2 - 4u^2v + 8v^2 = u^4 - 6u^2v + 9v^2 = (u^2 - 3v)^2.$$

Substituting back in for u, v we have $(x^2 + 2xy + y^2 - 3xy)^2 = (x^2 - xy + y^2)^2$.

Problem 4.5 Factor $x^3 + 3x^2 - 4$

Answer

$(x - 1)(x + 2)^2$

Solution

Note we can split up $-4 = -1 - 3$ so we have $x^3 - 1 + 3(x^2 - 1)$. Factoring gives

$$\begin{aligned}
(x - 1)(x^2 + x + 1) + 3(x - 1)(x + 1) &= (x - 1)(x^2 + x + 1 + 3x + 3) \\
&= (x - 1)(x^2 + 4x + 4) \\
&= (x - 1)(x + 2)^2
\end{aligned}$$

as our final answer.

Problem 4.6 Factor $(x^2 + 4x + 8)^2 + 3x(x^2 + 4x + 8) + 2x^2$.

Answer

$(x + 2)(x + 4)(x^2 + 5x + 8)$

Solution

Let $z = x^2 + 4x + 8$. Then our equation becomes $z^2 + 3xz + 2x^2$. Thinking of this as a quadratic in z, we can factor: $z^2 + 3xz + 2x^2 = (z + 2x)(z + x)$. Resubstituting we have

$$\begin{aligned}
(x^2 + 4x + 8 + 2x)(x^2 + 4x + 8 + x) &= (x^2 + 6x + 8)(x^2 + 5x + 8) \\
&= (x + 2)(x + 4)(x^2 + 5x + 8)
\end{aligned}$$

as our expression fully factored.

Problem 4.7 Factor $a^2 + (a + 1)^2 + (a^2 + a)^2$

Answer

$(a^2 + a + 1)^2$

Solution

Expand the first two terms, $a^2 + (a+1)^2 = a^2 + a^2 + 2a + 1 = 2(a^2 + a) + 1$, so $a^2 + (a+1)^2 + (a^2+a)^2 = (a^2+a)^2 + 2(a^2+a) + 1 = (a^2+a+1)^2$.

Problem 4.8 Factor the following.

(a) $2acx + 4bcx + adx + 2bdx + 4acy + 8bcy + 2ady + 4bdy$

Answer

$(a+2b)(2c+d)(x+2y)$

Solution

Grouping,

$$
\begin{aligned}
& (2acx + 4bcx) + (adx + 2bdx) + (4acy + 8bcy) + (2ady + 4bdy) \\
= \ & 2cx(a+2b) + dx(a+2b) + 4cy(a+2b) + 2dy(a+2b) \\
= \ & (a+2b)(2cx + dx + 4cy + 2dy) \\
= \ & (a+2b)((2c+d)x + (2c+d)(2y)) \\
= \ & (a+2b)(2c+d)(x+2y)
\end{aligned}
$$

(b) $1 + 2a + 3a^2 + 4a^3 + 5a^4 + 6a^5 + 5a^6 + 4a^7 + 3a^8 + 2a^9 + a^{10}$.

Answer

$(a+1)^2(a^2+a+1)^2(a^2-a+1)^2$.

Solution

Grouping, this equals $(1 + a + a^2 + a^3 + a^4 + a^5)^2$, so answer is $(a+1)^2(a^2+a+1)^2(a^2-a+1)^2$.

Problem 4.9 Factor $a^5 + a + 1$.

Answer

$(a^2+a+1)(a^3-a^2+1)$

Solution

Add and minus a^2: $a^5 - a^2 + a^2 + a + 1 = a^2(a^3 - 1) + (a^2 + a + 1) = (a^2 + a + 1)(a^3 - a^2 + 1)$.

Problem 4.10 Evaluate the following: $\dfrac{(1994^2 - 2000)(1994^2 + 3985) \times 1995}{1991 \cdot 1993 \cdot 1995 \cdot 1997}$.

Answer

1996.

Solution

Let $x = 1994$, then the expression is

$$\frac{(x^2 - x - 6)(x^2 + 2x - 3)(x+1)}{(x-3)(x-1)(x+1)(x+3)} = \frac{(x+2)(x-3)(x+3)(x-1)(x+1)}{(x-3)(x-1)(x+1)(x+3)} = x+2 = 1996.$$

5 Solutions to Chapter 5 Examples

Problem 5.1 Find the real solutions to $(2x^2 - 3x + 1)^2 = 22x^2 - 33x + 1$.

Answer

$-3/2, 0, 3/2, 3$

Solution

Let $y = (2x^2 - 3x + 1)$. This gives $y^2 = 11y - 10$ so $y^2 - 11y + 10 = 0$ and $(y - 10)(y - 1) = 0$ so $y = 10, y = 1$. Setting $10, 1$ equal to $2x^2 - 3x + 1$ gives $x = -3/2, 0, 3/2, 3$ as our solutions.

Problem 5.2 Solve $(x^2 + x - 1)^2 + 2x(x^2 + x - 1) + x^2 = 4$.

Answer

$-3, \pm 1$.

Solution

Let $y = x^2 + x - 1$ so we have $y^2 + 2xy + x^2 = 4$. Hence we get $(y + x)^2 - 4 = (y + x + 2)(y + x - 2) = 0$. Hence either $x^2 + 2x + 1 = (x + 1)^2 = 0$ so $x = -1$ or $x^2 + 2x - 3 = (x + 3)(x - 1) = 0$ so $x = -3$ or $x = 1$. Therefore the roots are $x = -3, \pm 1$.

Problem 5.3 Solve $3x^2 + (x + 2)^2 + (x^2 + x)^2 = 9$ over the reals.

Answer

$\pm -\dfrac{1}{2} \pm \dfrac{\sqrt{5}}{2}$

Solution

Expanding and combining the first two terms we have $3x^2 + (x + 2)^2 = 4x^2 + 4x + 4$. Hence if $z = x^2 + x$ we get $4z + 4 + z^2 = 9$ so $z^2 + 4z - 5 = (z + 5)(z - 1) = 0$. Hence either $x^2 + x + 5 = 0$ (no real roots as the discriminant is -19) or $x^2 + x - 1 = 0$ which has roots of $\pm -\dfrac{1}{2} \pm \dfrac{\sqrt{5}}{2}$ using the quadratic formula.

Problem 5.4 Solve the following:

(a) $\dfrac{15}{x+1} = \dfrac{15}{x} - \dfrac{1}{2}$.

Answer

$x = -6$ and $x = 5$.

Solution

Multiplying $2x(x+1)$ we get $30x = 30(x+1) - (x(x+1))$ so $30x = 30x + 30 - x^2 - x$ or $x^2 + x - 30 = 0$. Therefore $(x+6)(x-5) = 0$ so $x = -6$ and $x = 5$ are solutions (double check neither are extraneous).

(b) $\dfrac{4x}{x^2-4} - \dfrac{2}{x-2} = \dfrac{x+1}{x+2}$.

Answer

$x = 1$

Solution

Multiplying $(x+2)(x-2)$ to get $4x - 2(x+2) = (x+1)(x-2)$, solve and get roots 1 and 2. 2 is extraneous, so $x = 1$.

Problem 5.5 Solve the following:

(a) $\dfrac{3-x}{2+x} = 5 - \dfrac{4(2+x)}{3-x}$.

Answer

$x = 1/2$ and $x = -1$.

Solution

Let $y = \dfrac{3-x}{2+x}$, solve and get $x = 1/2$ and $x = -1$.

(b) $\dfrac{x-3}{x+1} - \dfrac{x+1}{3-x} = \dfrac{5}{2}$.

Answer

$x = 7$ and $x = -5$.

Solution

Let $y = \dfrac{x-3}{x+1}$. Then $x = 7$ and $x = -5$. Both are roots after verifying.

Problem 5.6 Solve the equation $\dfrac{1}{2x^2-3} - 8x^2 + 12 = 0$.

Answer

$x = \pm\dfrac{\sqrt{7}}{2}$ and $x = \pm\dfrac{\sqrt{5}}{2}$.

Solution

Let $y = 2x^2 - 3$, then $\dfrac{1}{y} - 4y = 0$. Solve and get $y = \pm\dfrac{1}{2}$, and solve for x, get $x = \pm\dfrac{\sqrt{7}}{2}$ and $x = \pm\dfrac{\sqrt{5}}{2}$, all are verified to be solutions.

Problem 5.7 Solve: $\left(\dfrac{x+1}{x^2-1}\right)^2 - 4\left(\dfrac{x+1}{x^2-1}\right) + 3 = 0$.

Answer

$x = 2$ and $x = 4/3$.

Solution

$x = -1$ is not a root, so cancel it and simplify. Then let $y = \dfrac{1}{x-1}$. Solve and get $x = 2$ and $x = 4/3$. Both are roots after verifying.

Problem 5.8 Solve $2x^4 - 9x^3 + 14x^2 - 9x + 2 = 0$.

Answer

$x = 1$, $x = 2$, and $x = 1/2$.

Solution

Divide everything by x^2: $2\left(x^2 + \dfrac{1}{x^2}\right) - 9\left(x + \dfrac{1}{x}\right) + 14 = 0$. Let $y = x + \dfrac{1}{x}$, then

$y^2 = x^2 + \dfrac{1}{x^2} + 2$, so $2(y^2 - 2) - 9y + 14 = 0$, and $y = 2$ and $y = 5/2$. So $x = 1$, $x = 2$, and $x = 1/2$.

Problem 5.9 Solve $(x^2 - 1)(x^2 + 10x + 24) = 24$.

Answer

$-3, -2, -\dfrac{5}{2} \pm \dfrac{\sqrt{57}}{2}$

Solution

Factoring we have $(x^2 - 1)(x^2 + 10x + 24) = (x - 1)(x + 1)(x + 4)(x + 6)$. Regrouping as $(x - 1)(x + 6)(x + 1)(x + 4)$ the LHS equals $(x^2 + 5x - 6)(x^2 + 5x + 4)$. Hence making the substitution $y = x^2 + 5x - 6$ we have $y(y + 10) = 24$ or $y^2 + 10y - 24 = 0$. Factoring we get $(y + 12)(y - 2) = 0$ so $y = 2$ or $y = -12$. Hence we have either $x^2 + 5x - 6 = 2$ which solving gives $x = -\dfrac{5}{2} \pm \dfrac{\sqrt{57}}{2}$ or $x^2 + 5x - 6 = -12$ which solving gives $x = -3, -2$. These are the four solutions.

Problem 5.10 Solve the equation $\dfrac{x-1}{x+1} + \dfrac{x-4}{x+4} = \dfrac{x-2}{x+2} + \dfrac{x-3}{x+3}$.

Answer

$x = 0, -5/2$.

Solution

Take away 1 from each fraction, then $\dfrac{2}{x+2} - \dfrac{1}{x+1} = \dfrac{4}{x+4} - \dfrac{3}{x+3}$, so $\dfrac{x}{x^2 + 3x + 2} = \dfrac{x}{x^2 + 7x + 12}$. So $x = 0$ or $x = -5/2$.

6 Solutions to Chapter 6 Examples

Problem 6.1 Graph the following equations. Where do they cross the x-axis?

(a) $y = \left| \dfrac{x}{2} - 2 \right| - 2$.

Answer

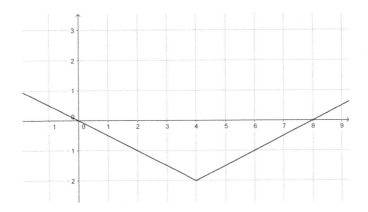

Solution

Note $x/2 - 2$ is positive when $x > 4$ and negative when $x < 4$. Hence $y = \dfrac{x}{2} - 4$ when $x \geq 4$ and $y = -\dfrac{x}{2}$ when $x < 4$. Combining these gives the graph above, which crosses the x-axis at 0 and 8.

(b) $y = |2x^2 + 5x - 12|$

Answer

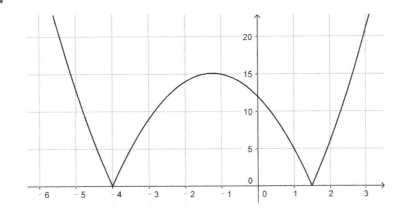

Solution

Note $2x^2 + 5x - 12 = (2x - 3)(x + 4)$ so $y = 2x^2 + 5x - 12$ is a parabola opening upwards with zeros at $x = -4$ and $x = 1.5$. Further, the vertex occurs when $x = \dfrac{-4 + 1.5}{2} = -1.25$ (and the y value at this point is -15.125. Hence the graph of $y = |2x^2 + 5x - 12|$ is the same when $x \leq -4$ and $x \geq 1.5$, but is mirrored across the x-axis in between. This gives the graph shown above.

Problem 6.2 Solve the following equations over the reals by considering cases.

(a) $|x| + 2 = |2x|$.

Answer

$x = \pm 2$.

Solution

Consider cases based on whether x is negative or not. If $x \geq 0$, then we have $x + 2 = 2x$ so $x = 2$. If $x < 0$, then we have $-x + 2 = -2x$ so $x = -2$. These are the two solutions.

(b) $|x^2 + 1| = 2|x - 1|$.

Answer

$x = -1 \pm \sqrt{2}$.

Solution

Note $|x^2 + 1| = x^2 + 1$ because squares are always positive. If $x \geq 1$ we have $x^2 + 1 = 2x - 2$ so $x^2 - 2x + 3 = 0$. Note the discriminant is negative, so this has no real roots. If $x < 1$ we have $x^2 + 1 = 2 - 2x$. Hence $x^2 + 2x - 1 = 0$ and we can solve for x to get $x = -1 \pm \sqrt{2}$ using the quadratic formula.

Problem 6.3 Solve the equation $|x| - 2 = -|1 - x|$.

Answer

$-\dfrac{1}{2}, \dfrac{3}{2}$.

Solution

Consider three cases: (i) $x < 0$, (ii) $0 \le x < 1$, or (iii) $x \ge 1$. In case (i) we have $-x - 2 = -1 + x$ so $2x = -1$ and $x = -1/2$. In case (ii), $x - 2 = -1 + x$ so $-2 = -1$ which is impossible. Lastly, for (iii), $x - 2 = -x + 1$ so $2x = 3$ and $x = 3/2$.

Problem 6.4 Solve $|x - |2x + 1|| = 3$.

Answer

$x = 2$ or $x = -4/3$

Solution

Case work: $x \ge -\frac{1}{2}$ or $x < -\frac{1}{2}$. If $x \ge -\frac{1}{2}$, $|x + 1| = 3$, so $x = 2$ or $x = -4$ (throw away -4). If $x < -\frac{1}{2}$, $|3x + 1| = 3$, then $x = 2/3$ (throw away) or $x = -4/3$. Final solution: $x = 2$ or $x = -4/3$.

Problem 6.5 Solve the following

(a) Solve the equation $|x^2 - 11x + 10| = |2x^2 + x - 45|$.

Answer

$x = -6 \pm \sqrt{91}$ and $x = \dfrac{5 \pm \sqrt{130}}{3}$.

Solution

It is a bit tedious to do case work on $x < 1$, $1 \le x < 9/2$, $9/2 \le x < 5$, $5 \le x < 10$ and $x \ge 10$. The faster way is to solve $x^2 - 11x + 10 = 2x^2 + x - 45$ and $x^2 - 11x + 10 = -(2x^2 + x - 45)$, and get $x = -6 \pm \sqrt{91}$ and $x = \dfrac{5 \pm \sqrt{130}}{3}$.

(b) $\dfrac{|x + 4|}{|x + 1|} = \dfrac{|x + 3|}{|x + 2|}$.

Answer

$\dfrac{-5}{2}, \dfrac{-5}{2} \pm \dfrac{\sqrt{3}}{2}$.

Solution

Regardless of how many of the terms inside the absolute values are negative, we either get (i) $\dfrac{x+4}{x+1} = \dfrac{x+3}{x+2}$ or (ii) $\dfrac{x+4}{x+1} = -\dfrac{x+3}{x+2}$. In case (i) we get

$$(x+4)(x+2) = (x+1)(x+3) \Rightarrow x^2+6x+8 = x^2+4x+3 \Rightarrow x = \frac{-5}{2}.$$

In case (ii) we get

$$(x+4)(x+2) = -(x+1)(x+3) \Rightarrow x^2+6x+8 = -x^2-4x-3 \Rightarrow 2x^2+10x+11 = 0 \Rightarrow x = \frac{-5}{2} \pm \frac{\sqrt{3}}{2},$$

using the quadratic equation. Double checking all 3 are solutions.

Problem 6.6 If $|m-2009| = -(n-2010)^2$, what is $(m-n)^{2011}$?

Answer

-1.

Solution

No squares are negative, and no absolute values are negative. So $m = 2009, n = 2010$, and $m-n = -1$. So $(m-n)^{2011} = -1$.

Problem 6.7 Solve the equation $|x^2+6x+1| = |(x+3)^2-4|$.

Answer

$-3 \pm \sqrt{6}$

Solution

Expanding the right side we get $|x+6x+5|$, so with the substitution $y = x^2+6x+1$ we get $|y| = |y+4|$. Hence we must have $y = -y-4$ so $y = -2$. Thus we need $x^2+6x+1 = -2$ or x^2+6x+3. The quadratic formula gives us $-3 \pm \sqrt{6}$ as roots.

Problem 6.8 The equation $|x^2-5x| = a$ has exactly two distinct real roots. What is the possible range of values for a?

Answer

$a = 0$ or $a > 25/4$.

Solution

$a = 0$ is obviously a good value. If $a > 0$, the two equations $x^2 - 5x - a = 0$ and $x^2 - 5x + a = 0$ combined have two distinct roots, that means the second equation has no real roots. Therefore, $5^2 + 4a > 0$ and $5^2 - 4a < 0$, which lead to $a > -\dfrac{25}{4}$ and $a > \dfrac{25}{4}$, thus $a > \dfrac{25}{4}$.

Problem 6.9 Find all solutions to $|||x + 1| - 1| - 1| = 1$.

Answer

$-4, -2, 0, 2$.

Solution

Case (1) If $||x + 1| - 1| - 1 = 1$, then $||x + 1| - 1| = 2$. Case (1a) $|x + 1| - 1 = 2$, $|x + 1| = 3$, so $x = 2$ or $x = -4$. Case (1b) $|x + 1| - 1 = -2$ impossible.
Case (2) If $||x + 1| - 1| - 1 = -1$, then $||x + 1| - 1| = 0$, so $|x + 1| = 1$, thus $x = 0$ or $x = -2$.
Therefore there are 4 solutions: $-4, -2, 0, 2$.

Problem 6.10 Consider the function $f(x) = |x^2 - 2x - 3| - |x^2 + x - 2|$.

(a) Break $f(x)$ into cases to write an equation for $f(x)$ without absolute values.

Answer

$$f(x) = \begin{cases} 2x^2 - x - 5 & \text{if } -2 \le x < -1 \\ 3x + 1 & \text{if } -1 \le x < 1 \\ -(2x^2 - x - 5) & \text{if } 1 \le x < 3 \\ -3x - 1 & \text{otherwise} \end{cases}$$

Solution

Factoring we have $x^2 - 2x - 3 = (x + 1)(x - 3)$ and $x^2 + x - 2 = (x + 2)(x - 1)$. Therefore $x^2 - 2x - 3 = -(x^2 - 2x - 3)$ when $-1 < x < 3$ and $x^2 - 2x - 3$ otherwise. Similarly $x^2 + x - 2 = -(x^2 + x - 2)$ when $-2 < x < 1$ and $x^2 + x - 2$ otherwise.

Therefore we consider four cases: (i) $x \ge 3$ or $x \le -2$, (ii) $1 \le x < 3$, (iii) $-1 \le x < 1$, and (iv) $-2 < x < -1$.

In case (i) we have $f(x) = x^2 - 2x - 3 - (x^2 + x - 2) = -3x - 1$. In case (ii) we have $f(x) = -(x^2 - 2x - 3) - (x^2 + x - 2) = -2x^2 + x + 5$. In case (iii) we have $f(x) = -(x^2 - 2x - 3) + (x^2 + x - 2) = 3x + 1$. Lastly, in case (iv) we have $f(x) = x^2 - 2x - 3 + (x^2 + x - 2) = 2x^2 - x - 5$. Hence

$$f(x) = \begin{cases} 2x^2 - x - 5 & \text{if } -2 \leq x < -1 \\ 3x + 1 & \text{if } -1 \leq x < 1 \\ -(2x^2 - x - 5) & \text{if } 1 \leq x < 3 \\ -3x - 1 & \text{otherwise} \end{cases}$$

(b) Find the domain, range, and any zeros of $y = f(x)$.

Answer

Domain and Range: All real numbers, Zeros: $-\dfrac{1}{3}, \dfrac{1 \pm \sqrt{41}}{4}$

Solution

Looking at a graph it is clear that the domain and range are all real numbers.

Note there are no zeros when $x < -2$ or $x > 3$. The linear term $3x + 1 = 0$ when $x = -1/3$ (which lies in the interval $-1 < x < 1$ as needed. Further, note zeros of $\pm(2x^2 - x - 5)$ are the same. Using the quadratic formula they are $-\dfrac{1}{3}, \dfrac{1 \pm \sqrt{41}}{4}$. Of these we see one lies in the range $-2 < x < -1$ and the other in $1 < x < 3$, so both are still zeros of $f(x)$. These are all the possible zeros as needed.

7 Solutions to Chapter 7 Examples

Problem 7.1 Find the domain and range of the following functions.

(a) $y = \sqrt{x^2 + 3x - 4}$.

Answer

Domain: $x \le -4$ or $x \ge 1$. Range: $y \ge 0$.

Solution

$x^2 + 3x - 4 = (x+4)(x-1)$ so this is non-negative when $x \le -4$ or $x \ge 1$. This gives the domain as we can take the square root of any non-negative number. Since $x^2 + 3x - 4$ can be any non-negative number, $y = \sqrt{x^2 + 3x - 4}$ can be any $y \ge 0$.

(b) Find the domain and range of $y = \sqrt{x^2 - 6x + 13}$.

Answer

Domain: All real numbers. Range: $y \ge 2$.

Solution

Completing the square we have $x^2 - 6x + 13 = x^2 - 6x + 9 + 4 = (x-3)^2 + 4$ so we have $y = \sqrt{(x-3)^2 + 4}$. Hence the function is always defined. Further, $(x-3)^2 \ge 0$, so $y \ge \sqrt{4} = 2$.

Problem 7.2 Find the real solutions to the following.

(a) $3 - \sqrt{2x - 3} = x$.

Answer

$x = 2$

Solution

$3 - x = \sqrt{2x - 3}$, so $9 - 6x + x^2 = 2x - 3$, and $x^2 - 8x + 12 = 0$, then $x = 2$ and $x = 6$. Check and find that $x = 2$ is the solution.

(b) $\sqrt{x+3} - \sqrt{3x-2} = -1$.

Answer

$x = 6$.

Solution

$\sqrt{x+3} = \sqrt{3x+2} - 1$, squaring, $x+3 = 3x-2 - 2\sqrt{3x-2} + 1$, thus $x - 2 = \sqrt{3x-2}$, square again, $x^2 - 7x + 6 = 0$, and get $x = 1$ or 6. $x = 1$ is extraneous. Therefore $x = 6$.

Problem 7.3 Solve: $\sqrt{x^2 + 3x + 7} - \sqrt{x^2 + 3x - 9} = 2$.

Answer

$x = 3$ and $x = -6$.

Solution

Let $y = \sqrt{x^2 + 3x + 7}$, then $y - \sqrt{y^2 - 16} = 2$, then $y^2 - 4y + 4 = y^2 - 16$, so $y = 5$. Solve for x to get $x = 3$ and $x = -6$, both check out alright.

Problem 7.4 Solve: $x^2 - \sqrt{3x^2 + 7} = 1$.

Answer

$x = \pm\sqrt{6}$.

Solution

Multiply 3 on both sides. Let $y = \sqrt{3x^2 + 7}$. Then $y = 5$, so $x = \pm\sqrt{6}$.

Problem 7.5 Solve: $\sqrt{\sqrt{x+4} + 4} = x$

Answer

$x = \dfrac{\sqrt{17} + 1}{2}$.

Solution

First note that $x > 0$. Square both sides: $\sqrt{x+4} + 4 = x^2$. Then $\sqrt{x+4} = x^2 - 4$, also note that $x > 2$. Squaring again to get $x + 4 = x^4 - 2 \cdot 4x^2 + 4^2$. To avoid factoring a quartic polynomial in x, we apply the following technique of "slave-as-master":

who says x must be the variable and 4 has to be the constant? Treat the constant 4 as a variable, and x as a parameter. Then it is a quadratic equation in the constant "4": $4^2 - (2x^2 + 1)4 + (x^4 - x) = 0$. Using the quadratic formula on 4, we get $4 = \dfrac{2x^2 + 1 \pm \sqrt{(2x^2 + 1)^2 - 4(x^4 - x)}}{2} = \dfrac{2x^2 + 1 \pm (2x + 1)}{2}$. So $4 = x^2 + x + 1$ or $4 = x^2 - x$. Solve and throw away the negative roots, $x = \dfrac{\sqrt{13} - 1}{2}$ or $x = \dfrac{\sqrt{17} + 1}{2}$. Further check shows that $x = \dfrac{\sqrt{13} - 1}{2} < 2$ is also extraneous. So there is only one root $x = \dfrac{\sqrt{17} + 1}{2}$. (It might help for easier understanding to do it this way: Let $y = 4$: $\sqrt{\sqrt{x + y} + y} = x$, and then solve for y in terms of x.)

Problem 7.6 Solve for real x: $\sqrt{\dfrac{x - 2}{x + 2}} + \sqrt{\dfrac{9x + 18}{x - 2}} = 4$.

Answer

$\dfrac{-5}{2}$.

Solution

Make the substitution $y = \sqrt{\dfrac{x - 2}{x + 2}}$ to get $y + \dfrac{3}{y} = 4$. Solving for y gives $y = 1, 3$. Hence $\dfrac{x - 2}{x + 2} = 1$ or 9. 1 is impossible, so $x - 2 = 9(x + 2)$ and hence $x = -5/2$.

Problem 7.7 Solve for x: $(x - \sqrt{3})x(x + 1) + 3 - \sqrt{3} = 0$.

Answer

$x = \sqrt{3} - 1$ and $x = \pm\sqrt[4]{3}$.

Solution

This is a cubic equation, and it is difficult to solve directly. So we let $y = \sqrt{3}$. Then $(x - y)x(x + 1) + y^2 - y = 0$. Expand and factor the left hand side: $x^3 - x^2y + x^2 - xy + y^2 - y = (x - y + 1)(x^2 - y) = 0$, so we have $x = y - 1$ and $x^2 = y$. Thus the solutions: $x = \sqrt{3} - 1$ and $x = \pm\sqrt[4]{3}$.

Problem 7.8 Let a be a real number, and the equation $x^2 + a^2x + a = 0$ has real roots for x. Find the maximum possible root x.

Answer

$x_{\max} = \sqrt[3]{2}/2.$

Solution

$x_{\max} = \sqrt[3]{2}/2.$ See the equation as a quadratic equation in a, $xa^2 + a + x^2 = 0$; the discriminant is $1^2 - 4x^3 \geq 0$, so $4x^3 \leq 1$, thus $x \leq \dfrac{1}{\sqrt[3]{4}} = \dfrac{\sqrt[3]{2}}{2}.$

Problem 7.9 Solve $\sqrt{2x+2} - \sqrt{x+3} = \sqrt{x+1} - \sqrt{2x+4}.$

Answer

$x = -1.$

Solution

Square both sides. After simplifying you get $\sqrt{(2x+2)(x+3)} = \sqrt{(x+1)(2x+4)}$ so $(2x+2)(x+3) = (x+1)(2x+4)$. Expanding and combining like terms we get $2x = -2$ so $x = -1$. Double checking, it is a solution.

Problem 7.10 For what range of k does $\sqrt{2x^2+4} = x + k$ have real solutions?

Answer

$k \geq \sqrt{2}.$

Solution

Squaring both sides we have $2x^2 + 4 = x^2 + 2kx + k^2$ so $x^2 - 2kx + 4 - k^2 = 0$. The discriminant is

$$\Delta = 4k^2 - 4(4 - k^2) = 8k^2 - 16.$$

Hence the equation has real roots when $8k^2 \geq 16$ so $k^2 \geq 2$, or $k \geq \sqrt{2}$ or $k \leq -\sqrt{2}$. However, if $k \leq -\sqrt{2}$ we have extraneous roots, so $k \geq \sqrt{2}$. (It may help to think of this part graphically.)

8 Solutions to Chapter 8 Examples

Problem 8.1 True or False. If the statement is false, explain how to correct the statement.

(a) If the degree of a polynomial $P(x)$ is d, then the number of terms of $P(x)$ is between 1 and d (inclusive).

Answer

False.

Solution

The number of terms is between 1 and $d+1$. For example, polynomials $2a^3$, $3x^3 - 2x$, $7t^3 + 5t^2 + 2t$, and $100u^3 + 4u^2 - 3u - 20$ are all polynomials of degree 3.

(b) If the degrees of polynomials $p(y)$ and $q(y)$ are d and e, then the degree of $p(y) \cdot q(y)$ is $d + e$.

Answer

True.

Solution

For example, $p(y) = x^2 + 2$ (degree 2) and $q(y) = x^3 - x$ (degree 3), then $p(y)q(y) = x^5 + x^3 - 2x$ (degree 5).

(c) If the degrees of polynomials $N(y)$ and $M(y)$ are d and e, then the degree of $N(M(y))$ is e^d.

Answer

False.

Solution

The degree is $d \cdot e$. For example, if $N(y) = x^2 + 2$ (degree 2) and $M(y) = x^3 - x$ (degree 3), then $N(M(y)) = (x^3 - x)^2 + 2 = x^6 - 2x^4 + x^2 + 2$ (degree 6).

Problem 8.2 For this problem, $f(x) = 3x + 2$, $g(x) = x - 7$, and $h(x) = x^2 - 4x + 4$.
Compute the following values:

(a) $f(g(4))$

Answer

-7.

(b) $g(g(g(g(g(35)))))$

Answer

0.

(c) $h(f(0))$

Answer

0.

(d) $h(f(100))$

Answer

90000.

(e) $f(g(1234567)) - g(f(1234567))$

Answer

-14.

Problem 8.3 In the polynomial $(7 + x)(1 + x^2)(5 + x^4)(2 + x^8)(3 + x^{16})(10 + x^{32})$,
what is the coefficient of x^{54}?

Answer

14.

Solution

Converted to binary, $54_{10} = 110110_2$. Thus the power $x^{54} = x^{32} \cdot x^{16} \cdot x^4 \cdot x^2$, and the coefficient is the product of the constant terms whose power of x does not appear, so the answer is $7 \cdot 2 = 14$.

Problem 8.4 Compute the quotient and remainders of the following:

(a) $(x^5 + 4x^4 + 4x^3 + 11x^2 + 16x + 6) \div (x^3 + 2x + 3)$.

Answer

Quotient: $x^2 + 4x + 2$, Remainder: 0.

Solution

Note that $(x^3 + 2x + 3)(x^2 + 4x + 2) = (x^5 + 4x^4 + 4x^3 + 11x^2 + 16x + 6)$.

(b) $(x^5 + 4x^4 + 4x^3 + 11x^2 + 16x + 6) \div (x^3 + 2x^2 + 5)$.

Answer

Quotient: $x^2 + 2x$, Remainder: $6x^2 + 6x + 6$.

Solution

Note that $(x^3 + 2x^2 + 5)(x^2 + 2x) + 6x^2 + 6x + 6 = (x^5 + 4x^4 + 4x^3 + 11x^2 + 16x + 6)$.

Problem 8.5 Solve $\dfrac{x^4 + 4x^3 + 2x^2 - 4x + 5}{x^2 + 2x - 1} = 4$.

Answer

$x = -3, 1$.

Solution

Dividing the numerator by the denominator using long division we have a quotient of $x^2 + 2x - 1$ and a remainder of 4, so the equation becomes:

$$x^2 + 2x - 1 + \frac{4}{x^2 + 2x - 1} = 4.$$

Substituting $y = x^2 + 2x - 1$ we can solve to get $y = 2$. Thus, $x^2 + 2x - 1 = 2$ so $x^2 + 2x - 3 = 0$. Factoring $x^2 + 2x - 3 = (x+3)(x-1)$ so $x = -3$ or $x = 1$.

Problem 8.6 Prove the Polynomial Remainder Theorem

Answer

Solution

By polynomial long division, we can write

$$P(x) = D(x)Q(x) + R(x),$$

where $\deg R(x) < \deg D(x)$. Since $\deg D(x) = 1$ ($x - a$ is linear), the remainder is degree) or a constant. So

$$P(x) = (x - a)Q(x) + r.$$

In the above identity, let $x = a$, then $P(a) = r$.

Problem 8.7 Solve $x^4 - 3x^3 - x^2 + 9x - 6 = 0$.

Answer

$x = 1, 2, \pm\sqrt{3}$.

Solution

Checking possible roots from the Rational Root Theorem, we see $1, 2$ are roots, so we can divide by $(x - 1)(x - 2) = x^2 - 3x + 2$. Using long division this gives us $x^2 - 3$, so the other roots are $\pm\sqrt{3}$.

Problem 8.8 Expand $(x^2 - x + 1)^6$ to get $a_{12}x^{12} + a_{11}x^{11} + \cdots + a_1x + a_0$. Find the value of $a_{12} + a_{10} + a_8 + a_6 + a_4 + a_2 + a_0$.

Answer

365.

Solution

Let $x = 1$:

$$1^6 = a_{12} + a_{11} + a_{10} + \cdots + a_2 + a_1 + a_0.$$

Let $x = -1$:

$$3^6 = a_{12} - a_{11} + a_{10} - \cdots + a_2 - a_1 + a_0.$$

Adding,
$$730 = 2(a_{12} + a_{10} + a_8 + a_6 + a_4 + a_2 + a_0),$$

So $a_{12} + a_{10} + a_8 + a_6 + a_4 + a_2 + a_0 = 365$.

Problem 8.9 Let x be a real number such that $x^3 + 4x = 8$. Determine the value of $x^7 + 64x^2$.

Answer

128.

Solution

Use $x^3 = -4x + 8$ to reduce the exponents. $x^7 + 64x^2 = x(-4x + 8)^2 + 64x^2 = 16x^3 - 64x^2 + 64x + 64x^2 = 16(x^3 + 4x) = 16 \times 8 = 128$.

Problem 8.10 Assume $(x - c)^2 \mid (4x^3 + 8x^2 - 11x + 3)$, find the value of c.

Answer

1/2.

Solution

Use the Rational Root Theorem to factor

$$4x^3 + 8x^2 - 11x + 3 = (2x - 1)^2(x + 3) = 4\left(x - \frac{1}{2}\right)^2 (x + 3).$$

9 Solutions to Chapter 9 Examples

Problem 9.1 Let $m \geq -1$ be a real number, and the equation $x^2 + 2(m-2)x + m^2 - 3m + 3 = 0$ has two distinct real roots x_1 and x_2. If $x_1^2 + x_2^2 = 6$, what is m?

Answer

$m = \dfrac{5 - \sqrt{17}}{2}$.

Solution

There are two distinct real roots, so $\Delta = 4(m-2)^2 - 4(m^2 - 3m + 3) = -4m + 4 > 0$. Therefore $m < 1$, and thus $-1 \leq m < 1$. So $6 = x_1^2 + x_2^2 = (x_1 + x_2)^2 - 2x_1x_2 = 4(m-2)^2 - 2(m^2 - 3m + 3) = 2m^2 - 10m + 10$. Solve for m we get $m = \dfrac{5 \pm \sqrt{17}}{2}$. Given the $-1 \leq m < 1$ range, we get $m = \dfrac{5 - \sqrt{17}}{2}$.

Problem 9.2 Let a, b, c, and d be the roots of $x^4 - 2x - 1990 = 0$. Find the value of $1/a + 1/b + 1/c + 1/d$.

Answer

$-1/995$.

Solution

$abc + bcd + cda + dab = 2$, and $abcd = -1990$, so $1/a + 1/b + 1/c + 1/d = (abc + bcd + cda + dab)/abcd = -1/995$.

Another way to solve this: Let $y = 1/x$, then $1 - 2y^3 - 1990y^4 = 0$, and the four roots for y are $1/a, 1/b, 1/c, 1/d$, so the answer is $-(-2)/(-1990) = -1/995$.

Problem 9.3 An $l \times w \times h$ rectangular box has surface area 38 and volume 12. If $l + w + h = 8$, find the dimensions of the box.

Answer

$4 \times 3 \times 1$.

Solution

Since $lwh = 12$ and $2(lw + wh + hl) = 38$, the values l, w, h are the three roots of polynomial equation $x^3 - 8x^2 + 19x - 12 = 0$. Factor the polynomial, $(x-1)(x-3)(x-4) = 0$, so the three roots are $1, 3, 4$. Thus it is a $4 \times 3 \times 1$ box.

Problem 9.4 Find the sum of the 17th powers of the 17 roots of $x^{17} - 3x + 1 = 0$.

Answer

-17.

Solution

Each of the roots satisfy $x_i^{17} - 3x_i + 1 = 0$, thus $x_i^{17} = 3x_i - 1$. Using Vieta's formula, $\sum x_i^{17} = \sum (3x_i - 1) = 3 \sum x_i - 17 = -17$.

Problem 9.5 Distinct real numbers a and b satisfies $(a+1)^2 = 3 - 3(a+1), 3(b+1) = 3 - (b+1)^2$. Find the value of $b\sqrt{\dfrac{b}{a}} + a\sqrt{\dfrac{a}{b}}$.

Answer

-23.

Solution

a and b are the roots of an equation $(x+1)^2 + 3(x+1) - 3 = 0$, which is $x^2 + 5x + 1 = 0$. The discriminant is positive. We have: $a + b = -5$, and $ab = 1$. Thus $b\sqrt{\dfrac{b}{a}} + a\sqrt{\dfrac{a}{b}} = -\dfrac{b}{a}\sqrt{ab} - \dfrac{a}{b}\sqrt{ab} = -\dfrac{a^2 + b^2}{ab}\sqrt{ab} = -\dfrac{(a+b)^2 - 2ab}{\sqrt{ab}} = -23$.

Problem 9.6 Find ordered pairs (x, y) of real numbers such that $x^2 - xy + y^2 = 13$ and $x - xy + y = -5$.

Answer

$(3, 4), (4, 3), (-2 + \sqrt{3}, -2 - \sqrt{3}), (-2 - \sqrt{3}, -2 + \sqrt{3})$.

Solution

Let $u = x + y$, $v = xy$, then $u^2 - 3v = 13$, and $u - v = -5$. Solve to get $(u, v) = (7, 12)$ or $(u, v) = (-4, 1)$. For each pair of (u, v), set up quadratic equation in t: $t^2 - ut + v = 0$ and x, y are the two roots. So $t^2 - 7t + 12 = 0$, or $t^2 + 4t + 1 = 0$. The solution for (x, y) are $(3, 4), (4, 3), (-2 + \sqrt{3}, -2 - \sqrt{3}), (-2 - \sqrt{3}, -2 + \sqrt{3})$.

Problem 9.7 If $x + y + z = 0$ and $x^3 + y^3 + z^3 = 288$, find the value of xyz.

Answer

96.

Solution

An important factoring formula: $x^3 + y^3 + z^3 - 3xyz = (x + y + z)(x^2 + y^2 + z^2 - xy - yz - zx)$. Then $288 - 3xyz = 0$, so $xyz = 96$.

Problem 9.8 The polynomial $p(x) = x^3 + 2x^2 - 5x + 1$ has three different roots a, b, and c. Find $a^3 + b^3 + c^3$.

Answer

−41.

Solution

$a + b + c = -2, ab + bc + ca = -5, abc = -1$. And $(a + b + c)^3 = a^3 + b^3 + c^3 + 3a^2b + 3ab^2 + 3b^2c + 3bc^2 + 3c^2a + 3ca^2 + 6abc = a^3 + b^3 + c^3 + 3(a + b + c)(ab + bc + ca) + abc$. So $a^3 + b^3 + c^3 = (-2)^3 - 3(-2)(-5) + (-1) = -41$.

There is more than one way to find the combination of the symmetric polynomials in a, b, c.

Problem 9.9 Let $x = \dfrac{2}{2 + \sqrt{3} - \sqrt{5}}, y = \dfrac{2}{2 + \sqrt{3} + \sqrt{5}}$, evaluate:

$$\frac{x^4 y^4}{x^4 + y^4 + 6x^2 y^2 + 4x^3 y + 4xy^3}.$$

Answer

$M = 97 - 56\sqrt{3}.$

Solution

Let $M = \dfrac{x^4 y^4}{x^4 + y^4 + 6x^2 y^2 + 4x^3 y + 4xy^3}$. Then

$$\frac{1}{M} = \frac{x^4 + y^4 + 6x^2 y^2 + 4x^3 y + 4xy^3}{x^4 y^4} = \frac{(x+y)^4}{x^4 y^4} = \left(\frac{1}{x} + \frac{1}{y}\right)^4.$$

Also $\dfrac{1}{x} + \dfrac{1}{y} = \dfrac{2 + \sqrt{3} - \sqrt{5}}{2} + \dfrac{2 + \sqrt{3} + \sqrt{5}}{2} = 2 + \sqrt{3}.$

Therefore $\dfrac{1}{M} = (2 + \sqrt{3})^4 = (7 + 4\sqrt{3})^2 = 97 + 56\sqrt{3}$, and then $M = 97 - 56\sqrt{3}.$

Problem 9.10 Let x and y be nonzero real numbers satisfying $|x| + y = 3$ and $|x|y + x^3 = 0$, Find the value of $x + y$.

Answer

$4 - \sqrt{13}.$

Solution

Substitute $y = 3 - |x|$ into $|x|y + x^3 = 0$, get $x^3 - x^2 + 3|x| = 0$. Case analysis for $x \geq 0$ and $x < 0$: If $x \geq 0$, $x^3 - x^2 + 3x = 0$, there are no nonzero real roots. If $x < 0$, $x^3 - x^2 - 3x = 0$, so $x = \dfrac{1 - \sqrt{13}}{2}$ (throw away the positive root), and $y = 3 + x$, so $x + y = 3 + 2x = 4 - \sqrt{13}.$

Printed in Great Britain
by Amazon

40410057R00077